BIRKHÄUSER

Progress in Mathematics
Volume 277

Timothy D. Browning

Quantitative Arithmetic of Projective Varieties

Birkhäuser
Basel · Boston · Berlin

Author:

Timothy D. Browning
School of Mathematics
University of Bristol
Bristol BS8 1TW
United Kingdom
e-mail: t.d.browning@bristol.ac.uk

2000 Mathematics Subject Classification: 11D45, 11G35, 11D72, 11E76, 11P55, 14G05, 14G10, 14G25

Library of Congress Control Number: 2009934090

Bibliographic information published by Die Deutsche Bibliothek.
Die Deutsche Bibliothek lists this publication in the Deutsche Nationalbibliografie; detailed bibliographic data is available in the Internet at http://dnb.ddb.de

ISBN 978-3-0346-0128-3 Birkhäuser Verlag AG, Basel · Boston · Berlin

Basel · Boston · Berlin
P.O. Box 133, CH-4010 Basel, Switzerland
Part of Springer Science+Business Media
Printed on acid-free paper produced from chlorine-free pulp. TCF ∞
Printed in Germany

ISBN 978-3-0346-0128-3 e-ISBN 978-3-0346-0129-0

9 8 7 6 5 4 3 2 1 www.birkhauser.ch

Ferran Sunyer i Balaguer (1912–1967) was a self-taught Catalan mathematician who, in spite of a serious physical disability, was very active in research in classical mathematical analysis, an area in which he acquired international recognition. His heirs created the Fundació Ferran Sunyer i Balaguer inside the Institut d'Estudis Catalans to honor the memory of Ferran Sunyer i Balaguer and to promote mathematical research.

Each year, the Fundació Ferran Sunyer i Balaguer and the Institut d'Estudis Catalans award an international research prize for a mathematical monograph of expository nature. The prize-winning monographs are published in this series. Details about the prize and the Fundació Ferran Sunyer i Balaguer can be found at

`http://www.crm.es/FSBPrize/ffsb.htm`

This book has been awarded the Ferran Sunyer i Balaguer 2009 prize.

INSTITVT D'ESTVDIS CATALANS MCMVII

The members of the scientific commitee of the 2009 prize were:

Hyman Bass
University of Michigan

Núria Fagella
Universitat de Barcelona

Joan Verdera
Universitat Autònoma de Barcelona

Alan Weinstein
University of California at Berkeley

Ferran Sunyer i Balaguer Prize winners since 1998:

1999 Patrick Dehornoy
Braids and Self-Distributivity, PM 192

2000 Juan-Pablo Ortega and Tudor Ratiu
Hamiltonian Singular Reduction, PM 222

2001 Martin Golubitsky and Ian Stewart
The Symmetry Perspective, PM 200

2002 André Unterberger
Automorphic Pseudodifferential Analysis and Higher Level Weyl Calculi, PM 209

Alexander Lubotzky and Dan Segal
Subgroup Growth, PM 212

2003 Fuensanta Andreu-Vaillo, Vincent Caselles and José M. Mazón
Parabolic Quasilinear Equations Minimizing Linear Growth Functionals, PM 223

2004 Guy David
Singular Sets of Minimizers for the Mumford-Shah Functional, PM 233

2005 Antonio Ambrosetti and Andrea Malchiodi
Perturbation Methods and Semilinear Elliptic Problems on R^n, PM 240

José Seade
On the Topology of Isolated Singularities in Analytic Spaces, PM 241

2006 Xiaonan Ma and George Marinescu
Holomorphic Morse Inequalities and Bergman Kernels, PM 254

2007 Rosa Miró-Roig
Determinantal Ideals, PM 264

2008 Luis Barreira
Dimension and Recurrence in Hyperbolic Dynamics, PM 272

To my wife Sinead

Contents

Preface

Over the millennia Diophantine equations have supplied an extremely fertile source of problems. Their study has illuminated ever increasing points of contact between very different subject areas, including algebraic geometry, mathematical logic, ergodic theory and analytic number theory. The focus of this book is on the interface of algebraic geometry with analytic number theory, with the basic aim being to highlight the rôle that analytic number theory has to play in the study of Diophantine equations.

Broadly speaking, analytic number theory can be characterised as a subject concerned with counting interesting objects. Thus, in the setting of Diophantine geometry, analytic number theory is especially suited to questions concerning the "distribution" of integral and rational points on algebraic varieties. Determining the arithmetic of affine varieties, both qualitatively and quantitatively, is much more complicated than for projective varieties. Given the breadth of the domain and the inherent difficulties involved, this book is therefore dedicated to an exploration of the projective setting.

This book is based on a short graduate course given by the author at the I.C.T.P *School and Conference on Analytic Number Theory*, during the period 23rd April to 11th May, 2007. It is a pleasure to thank Professors Balasubramanian, Deshouillers and Kowalski for organising this meeting. Thanks are also due to Michael Harvey and Daniel Loughran for spotting several typographical errors in an earlier draft of this book. Over the years, the author has greatly benefited from discussing mathematics with Professors de la Bretèche, Colliot-Thélène, Fouvry, Hooley, Salberger, Swinnerton-Dyer and Wooley. A sincere debt of thanks is owed to them all. Finally, it is essential to single out Professor Heath-Brown for special gratitude, both as a mathematical inspiration and for the generosity of his explanations.

Chapter 1

Introduction

The study of integer solutions to Diophantine equations is a topic that is almost as old as mathematics itself. Since its inception at the hands of Diophantus of Alexandria in 250 A.D., it has been found to relate to virtually every mathematical field. Suppose that we are given a polynomial $f \in \mathbb{Z}[x_1, \ldots, x_n]$ and write

$$S_f := \{\mathbf{x} = (x_1, \ldots, x_n) \in \mathbb{Z}^n \setminus \{\mathbf{0}\} : f(\mathbf{x}) = 0\} \tag{1.1}$$

for the corresponding locus of non-zero integer solutions. There are a number of basic questions that can be asked about the set S_f.

- When is S_f non-empty?
- How large is S_f when it is non-empty?
- When S_f is infinite can we describe the set in some way?

Much of our progress has been driven by trying to understand the situation for equations in only $n = 2$ or 3 variables, with the arithmetic of curves being central in our understanding of Diophantine equations. The terrain for equations in 4 or more variables remains relatively obscure, however, with only a scattering of results and conjectures available.

The focus of this book will be on quantitative aspects of the arithmetic of higher-dimensional projective varieties. Thus our interest lies with the second and third questions posed above, for Diophantine equations $f = 0$ in which f is homogeneous and the corresponding zero locus S_f is infinite. The main goal is to understand how the counting function

$$N(f; B) := \#\{\mathbf{x} \in S_f : \|\mathbf{x}\| \leqslant B\} \tag{1.2}$$

behaves, as $B \to \infty$. Here $\|\cdot\| : \mathbb{R}^n \to \mathbb{R}_{>0}$ is an arbitrary choice of norm. We will always reserve $|\cdot|$ for the norm $|\mathbf{x}| := \max_{1 \leqslant i \leqslant n} |x_i|$, for any $\mathbf{x} \in \mathbb{R}^n$.

Aside from being intrinsically interesting in their own right, the study of functions like $N(f; B)$ often helps determine whether or not the equation $f = 0$

has any non-trivial integer solutions at all. In many applications of the Hardy–Littlewood circle method, for example, one is able to prove that S_f is infinite by showing that $N(f;B) \to \infty$ as $B \to \infty$. In addition to the solubility of Diophantine equations, there are a number of other situations where a proper understanding of $N(f;B)$ is extremely desirable. We will return to this topic in Section 1.3.

During the course of this work we will meet numerous estimates of one kind or another. It seems worthwhile recording some of the basic notation here. We will write $A(x) = O(B(x))$ to mean that there exists a constant $c > 0$ and $x_0 \in \mathbb{R}$ such that $|A(x)| \leqslant cB(x)$ for all $x \geqslant x_0$. We will often use the alternative notation $A(x) \ll B(x)$ or $B(x) \gg A(x)$. Furthermore, we will take $A(x) \asymp B(x)$ to mean $A(x) \ll B(x) \ll A(x)$ and $A(x) = o(B(x))$ to mean

$$\lim_{x\to\infty} \frac{A(x)}{B(x)} = 0.$$

Finally the relation $A(x) \sim B(x)$ will mean

$$\lim_{x\to\infty} \frac{A(x)}{B(x)} = 1.$$

The implied constants in our work will be uniform unless explicitly indicated otherwise by an appropriate subscript. We will occasionally find it convenient to depart from this convention, but such deviations will be clearly highlighted.

1.1 A naive heuristic

Given our discussion above, it is useful to have a general idea of which homogeneous polynomials f, hitherto called *forms*, might have an infinite zero locus S_f. Suppose that $f \in \mathbb{Z}[x_1,\dots,x_n]$ is a form of degree $d \geqslant 1$. Then for the vectors $\mathbf{x} \in \mathbb{Z}^n$ counted by $N(f;B)$, the values of $f(\mathbf{x})$ will all be of order B^d. In fact a positive proportion of them will have exact order B^d. Thus the probability that a randomly chosen value of $f(\mathbf{x})$ should vanish might be expected to be of order B^{-d}. Since the number of $\mathbf{x}$ to be considered has order B^n, this leads us to the following general expectation.

Heuristic. *When $n \geqslant d$ we have*

$$B^{n-d} \ll N(f;B) \ll B^{n-d}. \tag{1.3}$$

As a crude first approximation, therefore, this heuristic tells us that we might expect polynomials whose degree is less than the number of variables to have infinitely many solutions. Unfortunately there are a number of things that can conspire to upset this heuristic expectation. First and foremost, local conditions will often provide a reason for $N(f;B)$ to be identically zero no matter the values of d and n. By local obstructions we mean that the obvious necessary conditions for S_f to be non-empty fail. These are the conditions that the equation $f(\mathbf{x}) = 0$ should have a non-zero real solution $\mathbf{x} \in \mathbb{R}^n$, and secondly, that the congruence

$$f(\mathbf{x}) \equiv 0 \pmod{p^k}$$

should be soluble for every prime power p^k, with $p \nmid \mathbf{x}$.

It is quite easy to construct examples that illustrate the failure of these local conditions. For example, the equation

$$x_1^{2d} + \cdots + x_n^{2d} = 0$$

does not have any integer solutions, since it patently does not have any real solutions. Let us now exhibit an example, due to Mordell [94], of a polynomial equation that fails to have integer solutions because it fails to have solutions as a congruence modulo prime powers. Let K be a number field of degree d over $\mathbb{Q}$, with ring of integers $\mathscr{O}_K$, such that the rational prime p is inert in $\mathscr{O}_K$. Write

$$\mathbf{N}(y_1, \ldots, y_d) := N_{K/\mathbb{Q}}(y_1\omega_1 + \cdots + y_d\omega_d)$$

for the corresponding norm form, where $\omega_1, \ldots, \omega_d$ is a basis for K over $\mathbb{Q}$. Then $\mathbf{N}$ is a homogeneous polynomial of degree d, with coefficients in $\mathbb{Z}$. Exercise 1.1 shows that $p \mid \mathbf{N}(\mathbf{y})$ if and only if $p \mid \mathbf{y}$, for any $\mathbf{y} \in \mathbb{Z}^d$. We define the form

$$f_1 := \mathbf{N}(x_1, \ldots, x_d) + p\mathbf{N}(x_{d+1}, \ldots, x_{2d}) + \cdots + p^{d-1}\mathbf{N}(x_{d^2-d+1}, \ldots, x_{d^2}), \quad (1.4)$$

which has degree d and d^2 variables. We claim that the only integer solution to the equation $f_1(\mathbf{x}) = 0$ is the trivial solution $\mathbf{x} = \mathbf{0}$. To see this we argue by contradiction. Thus we suppose there to be a vector $\mathbf{x} \in \mathbb{Z}^{d^2}$ such that $f_1(\mathbf{x}) = 0$, with $\gcd(x_1, \ldots, x_{d^2}) = 1$. Viewed modulo p we deduce that $p \mid \mathbf{N}(x_1, \ldots, x_d)$, whence $p \mid (x_1, \ldots, x_d)$. Writing $x_i = py_i$ for $1 \leqslant i \leqslant d$, and substituting into the equation $f_1 = 0$, we find that

$$p^{d-1}\mathbf{N}(y_1, \ldots, y_d) + \mathbf{N}(x_{d+1}, \ldots, x_{2d}) + \cdots + p^{d-2}\mathbf{N}(x_{d^2-d+1}, \ldots, x_{d^2}) = 0.$$

But then we deduce in a similar fashion that $p \mid (x_{d+1}, \ldots, x_{2d})$. We may clearly continue in this fashion, ultimately concluding that $p \mid (x_1, \ldots, x_{d^2})$, which is a contradiction.

The polynomial (1.4) illustrates that for any d it is possible to construct examples of homogeneous polynomials in d^2 variables that have no non-zero integer solutions. The construction is purely local, relying upon showing that the polynomial fails to have a non-zero solution in $\mathbb{Q}_p^{d^2}$. It was conjectured by Artin that $\mathbb{Q}_p$ is a C_2 field, so that f should have a non-trivial p-adic zero as soon as $n > d^2$. The latter property is certainly true of forms of degree at most 3. However, Artin's conjecture is now known to be false, with Terjanian [118] having provided a counterexample with $p = 2, d = 4$ and $n = 18$. In a positive direction, Ax and Kochen [1] have used methods from mathematical logic to show that for every d

there is a number $p(d)$ such that f has a non-trivial p-adic zero provided $n > d^2$ and $p > p(d)$. When no restriction is placed on the size of p we know that there is a number v_d such that the form f has a non-trivial p-adic zero as soon as $n > v_d$. Brauer [8] achieved the first result in this direction using an elementary argument based on multiply nested inductions. The resulting value of v_d was too large to write down, but the central ideas have since been revisited and improved upon by Wooley [124], with the outcome that we may take $v_d \leqslant d^{2^d}$.

So far we have only seen examples of polynomials f for which the zero locus S_f is empty. In this case the corresponding counting function $N(f;B)$ is particularly easy to estimate! There are also examples which show that $N(f;B)$ may grow in quite unexpected ways, even when $n \geqslant d$. An equation that illustrates excessive growth is provided by the polynomial

$$f_2 := x_1^d - x_2(x_3^{d-1} + \cdots + x_n^{d-1}). \tag{1.5}$$

Here there are *trivial* solutions of the type $(0, 0, a_3, \ldots, a_n)$ which already contribute $\gg B^{n-2}$ to the counting function $N(f;B)$, whereas (1.3) predicts that we should have exponent $n - d$.

It is also possible to construct examples of varieties which demonstrate inferior growth, as observed by Wooley [125]. Let $n > d^2$ and choose any d^2 linear forms $L_1, \ldots, L_{d^2} \in \mathbb{Z}[x_1, \ldots, x_n]$ that are linearly independent over $\mathbb{Q}$. Consider the form

$$f_3 := f_1(L_1(x_1, \ldots, x_n), \ldots, L_{d^2}(x_1, \ldots, x_n)),$$

where f_1 is given by (1.4). Then it is clear that $N(f_3;B)$ has the same order of magnitude as the counting function associated to the system of linear forms $L_1 = \cdots = L_{d^2} = 0$. Since these forms are linearly independent we deduce that $N(f_3;B)$ has order of magnitude B^{n-d^2}, whereas (1.3) led us to expect an exponent $n - d$.

We have seen several reasons why (1.3) might fail — how about some evidence supporting it? One of the most outstanding achievements in this direction is the following very general result due to Birch [6].

Theorem 1.1. *Suppose $f \in \mathbb{Z}[x_1, \ldots, x_n]$ is a non-singular homogeneous polynomial of degree d in $n > (d-1)2^d$ variables. Assume that $f(\mathbf{x}) = 0$ has non-trivial solutions in $\mathbb{R}$ and each p-adic field $\mathbb{Q}_p$. Then there is a constant $c_f > 0$ such that*

$$N(f;B) \sim c_f B^{n-d},$$

as $B \to \infty$.

We will discuss the proof of this result for the case $d = 4$ in Section 8.2. Birch's result does not apply to either of the polynomials f_2, f_3 that we considered above, since both of these contain a rather large singular locus. Since generic homogeneous polynomials are non-singular, Birch's result answers our initial questions completely for typical forms with $n > (d-1)2^d$. It would be of considerable interest to reduce the lower bound for n, but except for $d \leqslant 4$ this has not been done. Theorem 1.1 is established using the Hardy–Littlewood circle method, and exhibits a common feature of all Diophantine problems successfully tackled via this machinery: the number of variables involved needs to be large compared to the degree. In particular, there is an obvious disparity between the range for n in Birch's result and the range for n in (1.3).

Before coming to the Hardy–Littlewood circle method, we will also discuss some of the other technology that has been brought to bear on the quantitative analysis of homogeneous Diophantine equations. Whereas the circle method is geared towards forms for which the number of variables n is large compared to the degree d, we will also meet machinery to deal with equations in which n is comparable in size with d, in addition to those equations for which n is much smaller than d.

1.2 The basic counting function

It turns out that phrasing things in terms of single homogeneous polynomial equations is far too restrictive. It is much more satisfactory to work with arbitrary projective algebraic varieties $V \subseteq \mathbb{P}^{n-1}$. All of the varieties that we will work with are assumed to be cut out by a finite system of homogeneous equations defined over $\mathbb{Q}$. Moreover, whenever we speak of a variety as being *irreducible* we will henceforth take this to mean that the variety is geometrically reduced and irreducible. In the case of varieties cut out by a single equation this is equivalent to the underlying polynomial being irreducible over the complex numbers.

Our main interest lies with those varieties V for which we expect the set $V(\mathbb{Q}) = V \cap \mathbb{P}^{n-1}(\mathbb{Q})$ to be infinite. Let $x = [\mathbf{x}] \in \mathbb{P}^{n-1}(\mathbb{Q})$ be a projective rational point, with $\mathbf{x} \in \mathbb{Z}^n$ chosen so that $\gcd(x_1, \ldots, x_n) = 1$. Then we define the *height* of x to be

$$H(x) := \|\mathbf{x}\|.$$

This therefore defines a function $H : \mathbb{P}^{n-1}(\mathbb{Q}) \to \mathbb{R}_{>0}$, and is none other than the exponential height function metrized by the choice of norm $\|\cdot\|$. Given any locally closed subset $U \subseteq V$, we may then define the counting function

$$N_U(B) := \#\{x \in U(\mathbb{Q}) : H(x) \leqslant B\}, \tag{1.6}$$

for each $B \geqslant 1$. All known examples of asymptotic formulae for the counting function $N_U(B)$ take the shape

$$N_U(B) \sim cB^a(\log B)^b,$$

as $B \to \infty$, for $a, b, c \geqslant 0$ such that $a \in \mathbb{Q}$ and $b \in \frac{1}{2}\mathbb{Z}$. In Chapter 2 we will encounter an attempt to interpret these quantities in terms of the underlying geometry of V.

The main difference between the counting function $N_U(B)$ and the quantity introduced in (1.2) is that we are now only interested in *primitive* integer solutions, by which we mean that the components of the vector $\mathbf{x} \in \mathbb{Z}^n$ should share no common prime factors. This formulation has the advantage of treating all scalar multiples of a given non-zero integer solution as a single point. We will henceforth write $\mathbb{Z}^n_{\mathrm{prim}}$ for the set of primitive vectors in $\mathbb{Z}^n$.

Recall the definition of the *Möbius function* $\mu : \mathbb{N} \to \{0, \pm 1\}$, which is given by

$$\mu(n) := \begin{cases} 0, & \text{if } p^2 \mid n \text{ for some prime } p, \\ 1, & \text{if } n = 1, \\ (-1)^r, & \text{if } n = p_1 \cdots p_r \text{ for distinct primes } p_1, \dots, p_r. \end{cases}$$

The Möbius function is a multiplicative arithmetic function that is of fundamental importance in analytic number theory. It is frequently engaged via the simple identity

$$\sum_{d \mid n} \mu(d) = \begin{cases} 1, & \text{if } n = 1, \\ 0, & \text{if } n \in \mathbb{Z}_{>1}. \end{cases}$$

It is through this rôle as a characteristic function that it figures in the quantitative study of Diophantine equations. We illustrate the procedure by showing how it allows us to relate the counting function (1.6) to our earlier counting function $N(f;B)$ in (1.2), when $U = V$ and $V \subset \mathbb{P}^{n-1}$ is a hypersurface with underlying form $f \in \mathbb{Z}[x_1, \dots, x_n]$. On noting that $\mathbf{x}$ and $-\mathbf{x}$ represent the same point in $\mathbb{P}^{n-1}$, it follows that

$$\begin{aligned} N_V(B) &= \frac{1}{2} \#\{\mathbf{x} \in \mathbb{Z}^n_{\mathrm{prim}} : f(\mathbf{x}) = 0, \ \|\mathbf{x}\| \leqslant B\} \\ &= \frac{1}{2} \sum_{k=1}^{\infty} \mu(k) \#\{\mathbf{x} \in \mathbb{Z}^n : f(\mathbf{x}) = 0, \ k \mid \mathbf{x}, \ \|\mathbf{x}\| \leqslant B\}. \end{aligned}$$

But then a simple change of variables furnishes

$$N_V(B) = \frac{1}{2} \sum_{k=1}^{\infty} \mu(k) N(f; k^{-1}B). \tag{1.7}$$

This process of using the Möbius function will henceforth be termed *Möbius inversion.*

The simplest sort of subvariety in $\mathbb{P}^{n-1}$ is obtained by taking f to be identically zero. This corresponds to taking $V = \mathbb{P}^{n-1}$. Schanuel [107] has obtained an asymptotic formula for $N_{\mathbb{P}^{n-1}}(B)$. There is a natural way to define a height function on $\mathbb{P}^{n-1}(K)$ for any algebraic number field K, and it is to this more general context that Schanuel's result applies. It will be instructive to present a proof of this result in the case $K = \mathbb{Q}$.

Theorem 1.2. *Let $n \geqslant 2$ and let $H : \mathbb{P}^{n-1}(\mathbb{Q}) \to \mathbb{R}_{>0}$ be the height function metrized by the choice of norm $|\mathbf{z}| = \max_{1\leqslant i\leqslant n} |z_i|$. Then we have*

$$N_{\mathbb{P}^{n-1}}(B) = \frac{2^{n-1}}{\zeta(n)} B^n + O_n\big(B^{n-1}(\log B)^{b_n}\big),$$

where

$$b_n := \begin{cases} 1, & \text{if } n = 2, \\ 0, & \text{if } n > 2. \end{cases}$$

Proof. Let $n \geqslant 2$. Our starting point is (1.7), which with f identically zero gives

$$N_{\mathbb{P}^{n-1}}(B) = \frac{1}{2}\sum_{k\leqslant B} \mu(k)\#\{\mathbf{x} \in \mathbb{Z}^n \setminus \{\mathbf{0}\} : |\mathbf{x}| \leqslant k^{-1}B\}.$$

If $[X]$ denotes the integer part of any real number $X > 0$, then it is clear that there are precisely $2[X] + 1$ integers n such that $|n| \leqslant X$. Hence it follows that

$$N_{\mathbb{P}^{n-1}}(B) = \frac{1}{2}\sum_{k\leqslant B} \mu(k)\big((2[k^{-1}B] + 1)^n - 1\big).$$

We now apply the basic estimate $[X] = X + O(1)$, which gives

$$\begin{aligned} N_{\mathbb{P}^{n-1}}(B) &= \frac{1}{2}\sum_{k\leqslant B} \mu(k)\Big(\frac{2^n B^n}{k^n} + O_n\Big(\frac{B^{n-1}}{k^{n-1}}\Big)\Big) \\ &= 2^{n-1}B^n \sum_{k\leqslant B} \frac{\mu(k)}{k^n} + O_n\Big(B^{n-1}\sum_{k\leqslant B} \frac{1}{k^{n-1}}\Big). \end{aligned}$$

We would now like to extend the sum over $k \leqslant B$ in the main term to a sum over all $k \in \mathbb{N}$. The error in doing this is at most

$$2^{n-1}B^n \sum_{k>B} \frac{1}{k^n} \leqslant 2^{n-1}B^n \int_B^\infty \frac{1}{t^n}\mathrm{d}t \ll_n B^{n-1}.$$

On noting that $\sum_{k\leqslant B} k^{1-n} \ll_n (\log B)^{b_n}$, with b_n as in the statement of the theorem, it easily follows that

$$N_{\mathbb{P}^{n-1}}(B) = 2^{n-1}B^n \sum_{k=1}^{\infty} \frac{\mu(k)}{k^n} + O_n\big(B^{n-1}(\log B)^{b_n}\big).$$

This completes the proof of Theorem 1.2. □

It follows from Theorem 1.2 that the counting function $N_U(B)$ in (1.6) is bounded for each B, no matter what the choice of U and V. In Figure 1.1 we have presented a pictorial representation of the set

$$\big\{[x_1, x_2, 1] \in \mathbb{P}^2(\mathbb{Q}) : \ H([x_1, x_2, 1]) \leqslant 30\big\}.$$

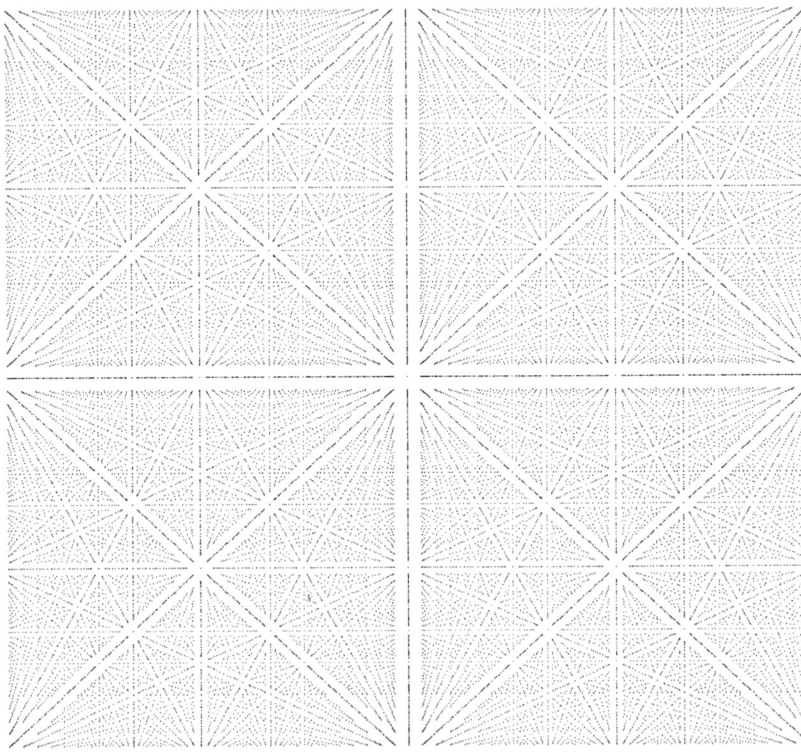

Figure 1.1: Points in $\mathbb{P}^2(\mathbb{Q})$ of height $\leqslant 30$

It indicates that one cannot hope to expect complete uniformity in the distribution of rational points, even in the simplest possible case.

1.3 Influence of analytic number theory

Much of this book is geared towards showing how tools rooted in analytic number theory can be used to study quantitative problems in Diophantine geometry. However there is a very real sense in which the converse is also true. That is to say, there are many problems in analytic number theory which can be recast in terms of counting rational or integral points on algebraic varieties constrained to (possibly quite lopsided) regions, some of which require applications of Diophantine geometry to successfully analyse. In this discussion we will restrict ourselves to those problems in analytic number theory that are linked with counting problems for projective varieties.

1.3.1 Paucity results

Following the rise to fame of Fermat's "last theorem", there was something of an industry in producing other equations which one could claim to have only trivial solutions in integers. One of the major players here was Euler who conjectured in 1769 that no non-negative kth power can be written as the sum of $k-1$ non-negative kth powers. This is equivalent to the equation

$$x_1^k + \cdots + x_{k-1}^k = x_k^k$$

possessing only trivial solutions in $\mathbb{Z}_{\geqslant 0}^k$ for every $k \geqslant 3$, where trivial means $x_k = x_i$ for some $1 \leqslant i \leqslant k-1$ and all other components zero. The case $k=3$ recovers the cubic case of Fermat's equation. This is now known to be false in general, however, with Elkies [56] refuting the case $k=4$, the smallest counter-example being the solution

$$95800^4 + 217519^4 + 414560^4 = 422481^4.$$

Lander and Parkin [87] handled the case $k=5$ somewhat earlier, by finding the solution

$$27^5 + 84^5 + 110^5 + 133^5 = 144^5.$$

Given that Elkies finds infinitely many counter-examples to Euler's conjecture in the case $k=4$, it would be interesting to find an infinite family of solutions for $k=5$ too.

Let $\nu_d(x)$ denote the number of positive integers not exceeding x which can be written as a sum of two dth powers of non-negative integers, in essentially more than one way. It is a long-standing conjecture that $\nu_d(x)=0$ for $d \geqslant 5$, and in the absence of its resolution, it is desirable to find as keen an upper bound for $\nu_d(x)$ as possible. Here, the upper bound $\nu_d(x) \ll_{d,\varepsilon} x^{\frac{2}{d}+\varepsilon}$ is completely trivial, and follows from divisor function estimates. Once cast in terms of Diophantine equations the above conjecture predicts that for $d \geqslant 5$ the only positive integer solutions to the equation

$$x_1^d + x_2^d = x_3^d + x_4^d \tag{1.8}$$

satisfy $\{x_1, x_2\} = \{x_3, x_4\}$. We call these the *trivial solutions*. When $d=3$ or 4 it is well-known that there are infinitely many non-trivial solutions. In general it is clear that one can get information about $\nu_d(x)$ by studying $N_U(x^{\frac{1}{d}})$, where U is the open subset of the surface (1.8), in which the coordinates satisfy $\{|x_1|, |x_2|\} \neq \{|x_3|, |x_4|\}$. The topic of estimating $\nu_d(x)$ has been extensively pursued in the literature. To illustrate this fact we have collected together some of the significant milestones in its study in Table 1.1. The final entry in the table still constitutes the best result in this direction.

$\nu_d(x) \ll_{d,\varepsilon}$	Who?
$x^{\frac{2}{d}+\varepsilon}$	trivial
$x^{\frac{5}{3d}+\varepsilon}$	Hooley [81, 82]
$x^{\frac{3}{2d}+\frac{1}{d(d-1)}+\varepsilon}$	Skinner & Wooley [113]
$x^{\frac{2}{3d}+\varepsilon} + x^{\frac{3}{d\sqrt{d}}+\frac{2}{d(d-1)}+\varepsilon}$	Browning [21]
$x^{\frac{3}{2d}+\frac{1}{d(2d-2)}+\varepsilon}$	Browning & Heath-Brown [31]
$x^{\frac{3}{d\sqrt{d}}+\varepsilon}$	Salberger [106]

Table 1.1: Estimates for $\nu_d(x)$

1.3.2 Waring's problem

It is a famous result of Lagrange from 1770 that every positive integer can be written as the sum of 4 squares of non-negative integers. This is the least number of squares with which one can hope to achieve such a representation, since we know from work of Legendre that a positive integer is the sum of 3 squares if and only if it is not of the form $4^i(8j+7)$ for $i, j \in \mathbb{Z}_{\geqslant 0}$. *Waring's problem* is concerned with producing an upper bound for the function $G(k)$, defined for $k \in \mathbb{N}$ as the least number of kth powers needed to represent all sufficiently large positive integers. Thus one has $G(2) = 4$, as follows from our discussion above.

The first explicit upper bound for $G(k)$ came through the pioneering work of Hardy and Littlewood [64] in 1922. This gave

$$G(k) \leqslant (k-1)2^{k-1} + 5,$$

and laid down the foundations of the Hardy–Littlewood circle method at the same time. Needless to say, there has been a great deal of activity in reducing the lower bound for $G(k)$ in the intervening years. A comprehensive discussion of Waring's problem can be found in the survey by Vaughan and Wooley [120]. In general, for large k, the best bound that we have is

$$G(k) \leqslant k(\log k + \log\log k + O_k(1)),$$

which is due to Wooley [122]. In approaches to Waring's problem using the circle method one meets the equation

$$x_1^k + \cdots + x_s^k = x_{s+1}^k + \cdots + x_{2s}^k. \tag{1.9}$$

It turns out that if one can succeed in showing that there are $O_{k,s,\varepsilon}(B^{2s-k+\varepsilon})$ solutions in the range $0 \leqslant x_1, \ldots, x_{2s} \leqslant B$, for any fixed $\varepsilon > 0$, then one can deduce that

$$G(k) \leqslant \max\{2k+1, \Gamma_0(k)\},$$

where $\Gamma_0(k)$ is there to ensure local solubility and is the least integer s such that the equation $x_1^k + \cdots + x_s^k = n$ always has a non-trivial solution in $\mathbb{Q}_p^s$. It is interesting to note that the upper bound $O_{k,s,\varepsilon}(B^{2s-k+\varepsilon})$ is predicted by the Manin conjecture discussed in the following chapter, at least for $s > \frac{k}{2}$.

Let $r_k(n)$ denote the number of representations of a positive integer n as the sum of k non-negative kth powers. As discussed by Hooley [83], a potentially useful instrument in the study of Waring's problem is *Hypothesis* K^*. This asserts a bound of the form

$$S_k(X) := \sum_{n \leqslant X} r_k(n)^2 = O_{k,\varepsilon}(X^{1+\varepsilon}),$$

for any $\varepsilon > 0$. This sum counts solutions of (1.9) with $s = k$ and $x_i^k \leqslant X$ for $1 \leqslant i \leqslant 2s$. Subject to the Riemann hypothesis for certain Hasse–Weil L-functions associated to cubic threefolds, this bound has been attained for the case $k = 3$ independently by Heath-Brown [72] and Hooley [83].

1.3.3 Vinogradov's mean value theorem

Consider the system of equations

$$x_1^j + \cdots + x_s^j = x_{s+1}^j + \cdots + x_{2s}^j, \quad (1 \leqslant j \leqslant k),$$

defining a projective variety $V_{k,s} \subset \mathbb{P}^{2s-1}$. *Vinogradov's mean value theorem* relates to the number $N_{k,s}(B)$, say, of integral solutions to this system of equations with $0 \leqslant x_1, \ldots, x_{2s} \leqslant B$.

By considering vectors $\mathbf{x} \in \mathbb{Z}^{2s}$ satisfying

$$\{x_1, \ldots, x_s\} = \{x_{s+1}, \ldots, x_{2s}\}, \tag{1.10}$$

it is easy to see that $N_{k,s}(B) \gg B^s$. Now let $\mathscr{H}$ denote the set of $\mathbf{h} \in \mathbb{Z}^k$ such that $|h_j| \ll B^j$ for $1 \leqslant j \leqslant k$. Furthermore, write $\mathscr{U}_{\mathbf{h}}(k,s)$ for the set of $\mathbf{x} \in \mathbb{Z}^{2s}$ for which

$$\sum_{i=1}^{s} (x_i^j - x_{s+i}^j) = h_j,$$

for $1 \leqslant j \leqslant k$, with $0 \leqslant x_1, \ldots, x_{2s} \leqslant B$. Then it follows that

$$B^{2s} \ll \sum_{\mathbf{h} \in \mathscr{H}} \#\mathscr{U}_{\mathbf{h}}(k,s) \ll \sum_{\mathbf{h} \in \mathscr{H}} N_{k,s}(B) \ll B^{\frac{k(k+1)}{2}} N_{k,s}(B).$$

Putting this together with our previous lower bound for $N_{k,s}(B)$, it therefore follows that

$$N_{k,s}(B) \gg \max\{B^s, B^{2s-\frac{k(k+1)}{2}}\},$$

for all $k, s \geqslant 1$.

Now if $s \leqslant k$, then all the solutions satisfy (1.10) and it follows that

$$N_{k,s}(B) \sim c_{k,s} B^s$$

in this case, for a suitable constant $c_{k,s} > 0$. Vaughan and Wooley [119] have shown that this asymptotic also holds for $s \leqslant k+1$. On the other hand, it follows from work of Wooley [123] that

$$N_{k,s}(B) \sim c'_{k,s} B^{2s-\frac{k(k+1)}{2}}$$

if $s \geqslant (1+o(1))k^2 \log k$, for a suitable constant $c'_{k,s} > 0$.

Bounds for $N_{k,s}(B)$ have numerous applications to analytic number theory, ranging from exponential sum estimates to information about the size of the Riemann zeta function on the critical line $\Re e(s) = \frac{1}{2}$. To go further it would be very useful to have a better idea of how $N_{k,s}(B)$ behaves for s of intermediate size in the above analysis. Returning to the notation $V_{k,s}$ for the corresponding variety, it seems likely that improved knowledge of the subvarieties of low degree in $V_{k,s}$ would be beneficial. However, even for particular choices of k and s, it seems to be a challenging problem in enumerative algebraic geometry to give a complete description of the various linear spaces that are permitted to lie in $V_{k,s}$.

1.3.4 Small solutions

Let $n \geqslant 3$ and let $f \in \mathbb{Z}[x_1, \ldots, x_n]$ be an indefinite form of degree $d \geqslant 2$, with coefficients of maximum modulus $\|f\|$. As indicated at the start of this chapter, a major influence in mathematics has been the pursuit of procedures for determining whether or not the set (1.1) is empty. One approach involves providing an effective upper bound for the smallest $\lambda \in \mathbb{N}$ with the property $S_f \neq \emptyset$ implies that there exists $\mathbf{x} \in S_f$ such that $|\mathbf{x}| = \max_{1 \leqslant i \leqslant n} |x_i| \leqslant \lambda$. Let us denote this quantity by $\Lambda_n(f)$ when it exists.

When $d = 2$, so that $f = Q$ is an indefinite quadratic form, there is a well-known result due to Cassels [38], which shows that

$$\Lambda_n(Q) \ll_n \|Q\|^{\frac{n-1}{2}}. \tag{1.11}$$

As an analytic number theorist one might be interested in the significance of the exponent $\frac{n-1}{2}$ and whether it can be substantially improved. In fact it was shown to be best possible by Kneser [39], via the ingenious example

$$Q_0(\mathbf{x}) := x_1^2 - (x_2 - cx_1)^2 - \cdots - (x_n - cx_{n-1})^2,$$

for any integer $c \geqslant 3$. It is self-evident that Q_0 is a non-singular indefinite quadratic form, with height $\|Q_0\| = 1 + c^2$. Moreover, we have $Q_0(\mathbf{a}) = 0$, where $\mathbf{a} = (1, c-1, c^2 - c, \ldots, c^{n-1} - c^{n-2})$. A little thought reveals that $\mathbf{a}$ is the unique solution to the equation $Q_0 = 0$, with least norm and positive first component. Noting that

$$|\mathbf{a}| = c^{n-1} - c^{n-2} > \frac{1}{2}c^{n-1} \gg \|Q_0\|^{\frac{n-1}{2}},$$

we deduce that (1.11) is indeed optimal.

It turns out, however, that Cassels' estimate for $\Lambda_n(Q)$ can be improved quite substantially for generic quadratic forms. This is the topic of recent work due to Browning and Dietmann [28]. Let $n \geqslant 5$ and let $Q \in \mathbb{Z}[x_1, \ldots, x_n]$ be a non-singular indefinite quadratic form of discriminant $\Delta \neq 0$, which is odd and square-free. Then it is shown that

$$\Lambda_n(Q) \ll_{\varepsilon,n} \|Q\|^{\frac{n-1}{2}} \left(\frac{\|Q\|^{n+\varepsilon}}{|\Delta|^{\frac{1}{2}(n-5-\alpha_n)}} \right)^{\frac{1}{n-3-\alpha_n}}, \tag{1.12}$$

for any $\varepsilon > 0$, where

$$\alpha_n = \begin{cases} 1, & \text{if } n \text{ is even,} \\ 0, & \text{if } n \text{ is odd.} \end{cases}$$

Suppose for the moment that $\theta \in \mathbb{R}$ is chosen so that $|\Delta| = \|Q\|^\theta$. Then we save something over Cassels' bound for $\Lambda_n(Q)$ as soon as

$$\theta > \frac{2n}{n-5-\alpha_n}.$$

As is well-known, the discriminant of a generic quadratic form has the same order of magnitude as the nth power of its height. Thus, for typical indefinite quadratic forms, one should be able to take $\theta = n$ in the above analysis. In this favourable setting we get an improvement over (1.11) as soon as $n \geqslant 9$.

The underlying idea in [28] is to produce an asymptotic formula for the weighted counting function

$$N_w(Q;B) := \sum_{\substack{\mathbf{x}\in\mathbb{Z}^n \\ Q(\mathbf{x})=0}} w(B^{-1}\mathbf{x}),$$

as $B \to \infty$, where $w : \mathbb{R}^n \to \mathbb{R}_{\geqslant 0}$ is a bounded function of compact support. The method of proof is based on the Hardy–Littlewood circle method. This is of course nothing new and many such asymptotic formulae are already available in the literature. However, for the purposes of proving (1.12), one needs an asymptotic formula which is completely uniform in the coefficients of Q. This eventually leads to an estimate of the shape

$$N_w(Q;B) = c_w(Q)B^{n-2} + \begin{cases} O_{\varepsilon,n}(\varpi_n(Q)B^{\frac{n}{2}+\varepsilon}), & \text{if } n \geqslant 6 \text{ is even,} \\ O_{\varepsilon,n}(\varpi_n(Q)B^{\frac{n-1}{2}+\varepsilon}), & \text{if } n \geqslant 5 \text{ is odd,} \end{cases}$$

for some explicit (albeit complicated) functions $c_w(Q), \varpi_n(Q)$ of the coefficients of Q. Comparing the main term and the error term one can therefore get precise bounds on the size of parameter B needed to ensure that $N_w(Q;B) \neq 0$, from which flows an effective bound for $\Lambda_n(Q)$.

The situation for forms of degree $d = 3$ is far less satisfactory. For general cubic forms $C \in \mathbb{Z}[x_1, \ldots, x_n]$ one of the the few results in the literature is due to

Pitman [101]. Let $\varepsilon > 0$. Then Pitman establishes the existence of a constant N_ε such that

$$\Lambda_n(C) \ll_{\varepsilon,n} \|C\|^{\frac{25}{6}+\varepsilon},$$

whenever $n \geqslant N_\varepsilon$. It is interesting to observe that the exponent of $\|C\|$ is independent of n, unlike the situation for quadratic forms discussed above. However, the number of variables needed to make this argument work is extremely large, due to the use of a diagonalisation process which reduces the problem to the easier one of bounding $\Lambda_n(C)$ for a diagonal cubic form. Browning, Dietmann and Elliot [35] have recently shown that $n \geqslant 17$ variables can be handled at the expense of replacing $\frac{25}{6}$ by 36×10^4 in Pitman's estimate. This is improved to 1071 when C is non-singular and $n = 17$. As highlighted by Swinnerton-Dyer [115, Problem 15], a real milestone in this domain would be a polynomial bound for $\Lambda_4(C)$ when C is a non-singular cubic form.

1.3.5 Divisor problems

In the following chapter we will discuss the Manin conjectures, a series of predictions concerning the density of rational points on suitable algebraic varieties. It turns out that a number of successful attacks upon the conjecture have made essential use of *divisor problems* for binary forms of one sort or another. The underlying problem here is one of pure analytic number theory and rests upon estimating sums of the shape

$$D(X;\varphi,F) := \sum_{a,b\leqslant X} \varphi(F(a,b)), \tag{1.13}$$

where $\varphi : \mathbb{N} \to \mathbb{R}_{\geqslant 0}$ is a suitable arithmetic function and $F \in \mathbb{Z}[x_1,x_2]$ is a binary form without repeated roots. Typically one insists that φ should be non-negative and multiplicative, satisfying the estimates $\varphi(n) \ll_\varepsilon n^\varepsilon$ for any $\varepsilon > 0$ and $\varphi(p^\ell) \leqslant A^\ell$ for all prime powers p^ℓ and a fixed constant $A > 0$.

If one is merely interested in upper bounds for $D(X;\varphi,F)$ then la Bretèche and Browning [11] have shown that the problem can be handled in a large degree of generality. A far more challenging problem arises in trying to produce asymptotic formulae for $D(X;\varphi,F)$ as $X \to \infty$. This problem has been most extensively studied in the case $\phi = \tau$, where $\tau(n) := \sum_{d|n} 1$ is the *divisor function*. Suppose that F has degree d and that it is irreducible over $\mathbb{Q}$. Handling forms of degree $d = 1$ or 2 is easy. Greaves [62] pioneered the study of higher degree forms by showing the existence of constants c_F, c'_F, with $c_F > 0$, such that

$$D(X;\tau,F) = c_F X^2 \log X + c'_F X^2 + O_{\varepsilon,F}(X^{\frac{27}{14}+\varepsilon}), \tag{1.14}$$

for any $\varepsilon > 0$, when $d = 3$. Greaves' proof uses exponential sums. It remained a significant open problem to deal with forms of degree 4 until Daniel [48] was able to show that

$$D(X;\tau,F) = c_F X^2 \log X + O_F(X^2 \log\log X),$$

for a constant $c_F > 0$, when $d = 4$. Daniel also sharpens Greaves' asymptotic formula, showing that the exponent in the error term in (1.14) may be replaced by $\frac{15}{8}$. Daniel's argument avoids using exponential sums and is based on some ideas from the geometry of numbers instead. Handling the divisor problem for quintic forms appears to be out of reach of current technology.

In Section 2.4 we will meet a few results which rely on the sort of technology introduced by Daniel to study divisor sums. In order to go further it seems very desirable to have estimates for $D(X;\tau,F)$ when F is a form of degree 3 or 4 which is not irreducible over the rationals, but which is still assumed to have no repeated roots. Preliminary steps in this direction have been made by la Bretèche and Browning [16], handling the case that F factorises over $\mathbb{Q}$ as $F = L_1L_2Q$, where L_1, L_2 are non-proportional binary linear forms and Q is an irreducible binary quadratic form. Furthermore, it seems that the rôle of divisor problems in counting problems is not yet fully explored and it may well be fruitful to consider the analogous situation for forms in more than 2 variables.

Exercises for Chapter 1

Exercise 1.1. Let K be a number field of degree d over $\mathbb{Q}$, with integral basis $\omega_1, \dots, \omega_d$. Let p be a rational prime such that the ideal (p) is prime in the ring of integers $\mathscr{O}_K$. For any $\alpha = y_1\omega_1 + \cdots + y_d\omega_d \in \mathscr{O}_K$, show that $p \mid N_{K/\mathbb{Q}}(\alpha)$ if and only if $p \mid \mathbf{y}$.

Exercise 1.2. Let $V \subset \mathbb{P}^{n-1}$ be a hypersurface defined by a non-singular form f of degree d, with $n > (d-1)2^d$. Mimicking the proof of Theorem 1.2, use Theorem 1.1 to deduce that

$$N_V(B) \sim \tilde{c}_f B^{n-d},$$

where $\tilde{c}_f = \frac{1}{2}\zeta(n-d)^{-1}c_f$.

Exercise 1.3. Establish the trivial bound $\nu_d(x) \ll_{d,\varepsilon} x^{\frac{2}{d}+\varepsilon}$ in Table 1.1.

Exercise 1.4. For $k \in \mathbb{N}$, let $g(k)$ be the least number of kth powers needed to represent every positive integer as a sum of kth powers. By considering the integer

$$n = 2^k\left[\left(\frac{3}{2}\right)^k\right] - 1,$$

deduce that $g(k) \geqslant 2^k + [(\frac{3}{2})^k] - 2$.

Chapter 2

The Manin conjectures

Around 1989 Manin [60] initiated a program to relate the asymptotic behaviour of counting functions to the intrinsic geometry of the underlying variety, for suitable families of algebraic varieties. As we will see, this subject affords a rich interplay of arithmetic and geometry.

Several of the varieties that we have looked at so far have many rational points, in the sense that $N_V(B)$ grows like a power of B. For such varieties it is natural to look at the quantity

$$\beta_V := \lim_{B\to\infty} \frac{\log N_V(B)}{\log B}, \tag{2.1}$$

assuming that this limit exists. In general we may consider β_U for any Zariski open subset $U \subseteq V$. It is clear that β_U gives a measure of "how large" the set $U(\mathbb{Q})$ is, since we will have

$$B^{\beta_U-\varepsilon} \ll N_U(B) \ll B^{\beta_U+\varepsilon}$$

for sufficiently large values of B and any $\varepsilon > 0$. We will henceforth refer to β_U as the *growth rate* of $U \subseteq V$. The insight of Manin was to try and relate β_U to the geometry of V via the introduction of a certain quantity $a(V)$. Before defining this quantity we will need to enhance our background with some facts from algebraic geometry.

2.1 Divisors on varieties

This book is not intended as a comprehensive introduction to algebraic geometry and so we will content ourselves with merely recalling the basic geometric objects and invariants that will play a rôle in our account of the Manin conjectures. For full details the reader is referred to Hartshorne [66]. We recall here our convention that a variety is said to be irreducible if it is geometrically reduced and irreducible.

2.1.1 The Picard group

Assume that $V \subset \mathbb{P}^{n-1}$ is non-singular. An irreducible codimension 1 subvariety of V is said to be a *prime divisor*. We let $\mathrm{Div}(V)$ be the free abelian group generated by finite formal sums of the shape

$$D = \sum n_Y Y,$$

with $n_Y \in \mathbb{Z}$ and Y running over prime divisors of V. This is the group of *Weil divisors* on V. A divisor $D \in \mathrm{Div}(V)$ is said to be *effective* if $n_Y \geqslant 0$ for all Y, and D is said to be *principal* if

$$D = \sum_Y \mathrm{ord}_Y(f) Y = D_f,$$

say, for some rational function $f \in \mathbb{C}(V)$. The intuitive idea behind the definition of the ord_Y function for a codimension 1 subvariety Y is that $\mathrm{ord}_Y(f) = k$ if f has a zero of order k along Y, while $\mathrm{ord}_Y(f) = -k$ if f has a pole of order k along Y. If f has neither a zero nor a pole along Y, then $\mathrm{ord}_Y(f) = 0$. Since $D_f + D_g = D_{fg}$ and $D_{\frac{1}{f}} = -D_f$, the principal divisors form a subgroup $\mathrm{PDiv}(V)$ of $\mathrm{Div}(V)$. We define the *geometric Picard group* associated to V to be

$$\mathrm{Pic}_{\overline{\mathbb{Q}}}(V) := \mathrm{Div}(V)/\,\mathrm{PDiv}(V).$$

This is just $\mathrm{Div}(V)$ modulo linear equivalence, where $D, D' \in \mathrm{Div}(V)$ are said to be *linearly equivalent* if $D - D'$ is principal. A divisor class $[D] \in \mathrm{Pic}_{\overline{\mathbb{Q}}}(V)$ is *effective* if there exists an effective divisor in the class. One may also construct the *geometric Néron–Severi group* $\mathrm{NS}_{\overline{\mathbb{Q}}}(V)$, which is $\mathrm{Div}(V)$ modulo a further equivalence relation called *algebraic equivalence.* When V is covered by curves of genus zero, as in many of the cases of interest to us, it turns out that $\mathrm{NS}_{\overline{\mathbb{Q}}}(V) = \mathrm{Pic}_{\overline{\mathbb{Q}}}(V)$. We illustrate the definition of $\mathrm{Pic}_{\overline{\mathbb{Q}}}(V)$ by calculating it in the simplest possible case $V = \mathbb{P}^{n-1}$.

Lemma 2.1. *We have* $\mathrm{Pic}_{\overline{\mathbb{Q}}}(\mathbb{P}^{n-1}) \cong \mathbb{Z}$.

Proof. A prime divisor on $\mathbb{P}^{n-1}$ has the form $Y = \{F = 0\}$ for some absolutely irreducible form $F \in \mathbb{C}[x_1, \dots, x_n]$. For such a divisor, define the degree of Y to be $\deg Y = \deg F$. Extend the definition of degree additively, so that

$$\deg\Big(\sum_Y n_Y Y\Big) = \sum_Y n_Y \deg Y.$$

The map $\deg : \mathrm{Div}(\mathbb{P}^{n-1}) \to \mathbb{Z}$ is clearly a homomorphism, and to establish the lemma it will suffice to show that the kernel of this map is precisely the subgroup $\mathrm{PDiv}(\mathbb{P}^{n-1})$. To see this, we note that $\deg D_f = 0$ for any rational function $f = \frac{F_1}{F_2}$. Indeed, the sum of the positive degree terms will be $\deg F_1$, whereas the sum of the negative degree terms will be $\deg F_2$, and these two degrees must coincide in order to have a well-defined rational function. Conversely, if $D = n_1Y_1 + \cdots + n_kY_k$ has degree zero, with $Y_i = \{F_i = 0\}$ for $1 \leqslant i \leqslant k$, then $f = F_1^{n_1} \cdots F_k^{n_k}$ is a well-defined rational function on $\mathbb{P}^{n-1}$ with $D_f = D$. This completes the proof of the lemma. $\square$

Returning to the setting of arbitrary non-singular varieties $V \subset \mathbb{P}^{n-1}$, we let

$$\Lambda_{\text{eff}}(V) := \{c_1[D_1] + \cdots + c_k[D_k] : c_i \in \mathbb{R}_{\geqslant 0},\ [D_i] \in \mathrm{NS}_{\overline{\mathbb{Q}}}(V) \text{ effective}\}. \quad (2.2)$$

This is the so-called *cone of effective divisors*.

2.1.2 The canonical divisor

We will write $K_V \in \mathrm{Div}(V)$ for the *canonical divisor*. This is a common abuse of notation: really K_V refers to the class of D_ω in $\mathrm{Pic}_{\overline{\mathbb{Q}}}(V)$ for any differential $(\dim V)$-form ω of V. It would take us too far afield to include a proper discussion of the canonical divisor, but this is readily supplied by consulting [66, Section II.8]. The negative $-K_V$ is called the *anticanonical divisor*.

2.1.3 The intersection form

For non-singular prime divisors which intersect transversally, the *intersection number* is simply the number of intersection points. Assume that V has dimension 2. As explained in [66, Section V.1], this notion can be extended to a symmetric bilinear form on $\mathrm{Div}(V)$. It induces the non-degenerate *intersection form* $(\cdot,\cdot)$ on $\mathrm{Pic}_{\overline{\mathbb{Q}}}(V)$. We let (D, D) be the *self-intersection number* of a divisor $D \in \mathrm{Div}(V)$.

2.1.4 Cubic surfaces

Let us illustrate the ideas that we have met so far by discussing some of the geometric properties of non-singular cubic surfaces $S \subset \mathbb{P}^3$. As demonstrated in the book by Manin [91], such surfaces are obtained by blowing up $\mathbb{P}^2$ along a collection of 6 points $P_1, \ldots, P_6$, which do not all lie on a conic in the plane and no 3 of which are collinear. For example, the familiar *Fermat cubic surface*

$$x_1^3 + x_2^3 + x_3^3 + x_4^3 = 0$$

is obtained by blowing up $\mathbb{P}^2$ along the six points

$$\begin{aligned} P_1 &= [1, -\rho^2, -\rho], & P_3 &= [0, 1, -\rho], & P_5 &= [-\rho, 1, 1],\\ P_2 &= [1, -\rho, -\rho^2], & P_4 &= [0, 1, -\rho^2], & P_6 &= [-\rho^2, 1, 1], \end{aligned}$$

where ρ is a primitive cube root of unity.

In general terms the *blow-up* of a point on a surface replaces the point in a particular way by an *exceptional divisor*, which is just a divisor with genus 0 and negative self-intersection number (see [66, Section I.4]). We will refer to a $(-k)$-*curve* as an exceptional divisor on a surface with self-intersection $-k$. As is well-known a non-singular cubic surface contains exactly 27 lines. These arise in the following way:

- there are 6 (-1)-curves E_i above P_i, for $1 \leqslant i \leqslant 6$;
- there are 15 strict transforms $L_{i,j}$ of the lines going through precisely 2 points P_i, P_j, for $1 \leqslant i < j \leqslant 6$;
- there are 6 strict transforms Q_i of the conics going through all but one of the 6 points.

If Λ is the strict transform of a line in $\mathbb{P}^2$ that does not go through any of the P_i, then a basis of the geometric Picard group $\mathrm{Pic}_{\overline{\mathbb{Q}}}(S)$ is given by

$$[\Lambda], [E_1], \ldots, [E_6].$$

These divisors satisfy the intersection behaviour

$$(\Lambda, \Lambda) = 1, \quad (\Lambda, E_i) = 0, \quad (E_i, E_j) = \begin{cases} -1, & \text{if } i = j, \\ 0, & \text{if } i \neq j. \end{cases}$$

The remaining divisors may be expressed in terms of these elements via the relations

$$[L_{i,j}] = [\Lambda] - [E_i] - [E_j], \quad [Q_i] = 2[\Lambda] - \sum_{j \neq i} [E_j]. \tag{2.3}$$

The adjunction formula implies that for any curve $C \subset S$ of genus g, one has the relation

$$(C, C + K_S) = 2g - 2.$$

It easily follows that the class of the anticanonical divisor $-K_S$ is given by

$$[-K_S] = 3[\Lambda] - \sum_{i=1}^{6} [E_j].$$

One can check that the hyperplane section has class $-3[\Lambda] + \sum_{i=1}^{6} [E_j]$ in $\mathrm{Pic}_{\overline{\mathbb{Q}}}(S)$, so that the cubic surface has very ample anticanonical divisor. When S is diagonal, it is not hard to write down the 27 lines explicitly. This will be carried out in Section 8.3.

2.2 The conjectures

Let $V \subset \mathbb{P}^{n-1}$ be a non-singular variety. Now let $H \in \mathrm{Div}(V)$ be a divisor corresponding to a hyperplane section. We define the real number

$$a(V) := \inf\{r \in \mathbb{R} : r[H] + [K_V] \in \Lambda_{\mathrm{eff}}(V)\},$$

where $\Lambda_{\text{eff}}(V)$ is the cone of effective divisors, defined in (2.2). The main thing to observe here is that $a(V)$ depends in an explicit way on the geometry of V over $\mathbb{C}$. Batyrev and Manin [2] were the first to introduce the quantity $a(V)$ in an effort to interpret β_V in (2.1), and it is their sequence of conjectures that we wish to present here. We begin with [2, Conjecture A].

Conjecture 2.1. *For all $\varepsilon > 0$ there exists a Zariski dense open subset $U \subseteq V$ such that $\beta_U \leqslant a(V) + \varepsilon$.*

Conjecture 2.1 deals with a very general class of projective algebraic varieties. For the remainder of this chapter we will focus our efforts on those varieties which stand the best chance of possessing many rational points, where we hope to say something much more precise.

A non-singular variety $V \subset \mathbb{P}^{n-1}$ is said to be *Fano* if K_V does not lie in the closure of the effective cone $\Lambda_{\text{eff}}(V) \subset \mathrm{NS}_{\overline{\mathbb{Q}}}(V) \otimes_{\mathbb{Z}} \mathbb{R}$. This is equivalent to $-K_V$ being ample, and implies in particular that V is covered by rational curves. As an example, suppose that V is a complete intersection, with $V = W_1 \cap \cdots \cap W_t$ for hypersurfaces $W_i \subset \mathbb{P}^{n-1}$ of degree d_i. Then V is Fano if and only if

$$d_1 + \cdots + d_t < n.$$

The following prediction, made in [2, Conjecture B], pins down the quantity in (2.1) more precisely.

Conjecture 2.2. *Assume that V is Fano and $V(\mathbb{Q})$ is Zariski dense in V. Then there exists a Zariski open subset $U \subseteq V$ such that $\beta_U = a(V)$.*

We have $a(V) = n - d_1 - \cdots - d_t$ when V is a non-singular complete intersection as above. In particular, when V is a hypersurface of degree d, we may deduce from Theorem 1.1 and Exercise 1.2 that Conjecture 2.2 holds with $U = V$ when n is sufficiently large in terms of d. It also holds for projective space. This follows from Theorem 1.2 and the fact that $[K_{\mathbb{P}^{n-1}}] = [-nH]$ in $\mathrm{Pic}_{\overline{\mathbb{Q}}}(\mathbb{P}^{n-1})$, whence $a(\mathbb{P}^{n-1}) = n$.

It would be too ambitious to give a comprehensive account of the distribution of rational points on Fano varieties. Thus we will content ourselves with concentrating attention on varieties with dimension at most 2, where there still remains much work to be done.

Let us summarise the situation for curves briefly. For simplicity we will discuss only projective plane curves $V \subset \mathbb{P}^2$ of degree d for which $V(\mathbb{Q}) \neq \emptyset$. There is a natural trichotomy among such curves, according to the genus g of the curve. For curves with $g = 0$, otherwise known as *rational curves*, it is possible to show that

$$N_V(B) \sim c_V B^{\frac{2}{d}}, \tag{2.4}$$

for an appropriate constant $c_V > 0$. This is in complete accordance with Conjecture 2.2. When $g = 1$, the curve is elliptic and it has been shown by Néron [112, Section 4.5] that

$$N_V(B) \sim c_V (\log B)^{\frac{r_V}{2}}, \tag{2.5}$$

where r_V denotes the rank of V. Thus although there can be infinitely many points in $V(\mathbb{Q})$, we see that the corresponding counting function grows much more slowly than for rational curves. Elliptic curves are not Fano, however, and so this is not covered by Conjecture 2.2. However, $a(V) = 0$ for curves of genus 1 and so (2.5) is in accordance with Conjecture 2.1. When $g \geqslant 2$, the work of Faltings [58] shows that $V(\mathbb{Q})$ is always finite, and so it does not a priori make sense to study $N_V(B)$.

We now discuss Fano varieties of dimension 2, beginning with some simple-minded numerics. Suppose that we are given a Fano variety V of dimension 2 and degree d, which is a non-singular complete intersection in $\mathbb{P}^{n-1}$. Thus $V = W_1 \cap \cdots \cap W_t$ for hypersurfaces $W_i \subset \mathbb{P}^{n-1}$ of degree d_i. To make life easier we assume that the intersection is transversal at a generic point of V. We are not interested in hyperplane sections of V, and so we will suppose without loss of generality that $d_i \geqslant 2$ for each $1 \leqslant i \leqslant t$. Then we have the following relations:

(i) $d_1 + \cdots + d_t < n$, [Fano]

(ii) $n - 1 - t = 2$, [complete intersection of dimension 2]

(iii) $d = d_1 \cdots d_t$, [Bézout's theorem]

(iv) $d_t \geqslant \cdots \geqslant d_1 \geqslant 2$.

It follows that the only possibilities are

$$(d; d_1, \ldots, d_t; n; t) \in \{(2; 2; 4; 1), (3; 3; 4; 1), (4; 2, 2; 5; 2)\}.$$

These surfaces correspond to a quadric in $\mathbb{P}^3$, a cubic surface in $\mathbb{P}^3$, and an intersection of 2 quadrics in $\mathbb{P}^4$, respectively. The arithmetic of quadrics is decidedly simpler and it is well-known that Conjecture 2.2 holds as soon as $n \geqslant 3$. Hence one would like to examine more closely the arithmetic of the latter two surfaces. In fact these are the two most familiar examples of *del Pezzo surfaces.*

We will see that not all del Pezzo surfaces are complete intersections, and so we have missed out on several surfaces in this analysis. Nonetheless, a significant part of this book will be dedicated to cubic surfaces in $\mathbb{P}^3$ and intersections of 2 quadrics in $\mathbb{P}^4$. It is now time to give a formal definition of a del Pezzo surface. Let us begin with a discussion of non-singular del Pezzo surfaces. Let $d \geqslant 3$. Then a *del Pezzo surface of degree d* is a non-singular surface $S \subset \mathbb{P}^d$ of degree d, with very ample anticanonical divisor $-K_S$. This latter condition is equivalent to the equality $[-K_S] = [H]$ in $\mathrm{Pic}_{\overline{\mathbb{Q}}}(S)$, for a hyperplane section $H \in \mathrm{Div}(S)$. The geometry of del Pezzo surfaces is very beautiful and well worth studying. However, to avoid straying from the main focus of this book, we will content ourselves with simply quoting the facts that are needed, referring the interested reader to the book by Manin [91].

It is well-known that del Pezzo surfaces $S \subset \mathbb{P}^d$ arise either as the quadratic Veronese embedding of a quadric in $\mathbb{P}^3$, which is a del Pezzo surface of degree 8 in $\mathbb{P}^8$ (isomorphic to $\mathbb{P}^1 \times \mathbb{P}^1$), or as the blow-up of $\mathbb{P}^2$ along $9 - d$ points in general position, in which case the degree of S satisfies

$$3 \leqslant d \leqslant 9.$$

Here $9 - d$ coplanar points are said to be in *general position* if no 3 of them are collinear and no 6 of them lie on a conic. One of the remarkable features of del Pezzo surfaces of small degree is that they contain finitely many lines. The precise number of lines is recorded in Table 2.1.

d	3	4	5	6
number of lines	27	16	10	6

Table 2.1: Lines on non-singular del Pezzo surfaces of degree d

For any non-singular del Pezzo surface $S \subset \mathbb{P}^d$ of degree d, the geometric Picard group $\mathrm{Pic}_{\overline{\mathbb{Q}}}(S)$ is a finitely generated free $\mathbb{Z}$-module, with

$$\mathrm{Pic}_{\overline{\mathbb{Q}}}(S) \cong \mathbb{Z}^{10-d}. \tag{2.6}$$

This is established in Manin [91], where an explicit basis for the group is also provided. It follows from the fact that S is the blow-up of $\mathbb{P}^2$ in $9 - d$ points. The example of cubic surfaces that we met in Section 2.1.4 is entirely typical of the general case. Let K be a splitting field for the finitely many lines contained in S. The final invariant that we will need to introduce is the *Picard group*

$$\mathrm{Pic}(S) := \mathrm{Pic}_{\overline{\mathbb{Q}}}(S)^{\mathrm{Gal}(K/\mathbb{Q})} \tag{2.7}$$

of the surface. This is just the set of elements in $\mathrm{Pic}_{\overline{\mathbb{Q}}}(S)$ that are fixed by the action of the Galois group $\mathrm{Gal}(K/\mathbb{Q})$.

It turns out that dealing with the arithmetic of del Pezzo surfaces of degree d gets easier as the degree increases. Since $[-K_S] = [H]$ in $\mathrm{Pic}_{\overline{\mathbb{Q}}}(S)$, we see that $a(S) = 1$ in Conjecture 2.2. In this setting there is a refinement of the prediction involving an asymptotic formula for the appropriate counting function. Write ρ_S for the rank of $\mathrm{Pic}(S)$ and let $U \subset S$ be the Zariski open subset formed by deleting the lines from S. Then we have the following [2, Conjecture C'].

Conjecture 2.3. *Suppose that $S \subset \mathbb{P}^d$ is a non-singular del Pezzo surface of degree d, with $S(\mathbb{Q}) \neq \emptyset$. Then there exists a constant $c_S > 0$ such that*

$$N_U(B) = c_S B (\log B)^{\rho_S - 1}\big(1 + o(1)\big). \tag{2.8}$$

We have already noted that the exponent of B agrees with Conjecture 2.2, since $a(S) = 1$. Moreover the exponent of $\log B$ is at most $9-d$, since the geometric Picard group has rank $10-d$. We will develop some heuristics to support this power of $\log B$ in Section 8.3. The value of the constant c_S has also received a conjectural interpretation at the hands of Peyre [96], and later also by Salberger [102].

Henceforth, unless explicitly indicated otherwise, we will assume that the height function H is metrized by the usual norm $|\mathbf{x}| = \max_{1\leqslant i\leqslant n}|x_i|$. There are a number of refinements of Conjecture 2.3 that are currently emerging. One such refinement is the following conjecture.

Conjecture 2.4. *Suppose that $S \subset \mathbb{P}^d$ is a non-singular del Pezzo surface of degree d, with $S(\mathbb{Q}) \neq \emptyset$. Then there exists a polynomial $P \in \mathbb{R}[x]$ of degree $\rho_S - 1$, and a real number $\delta > 0$, such that*

$$N_U(B) = BP(\log B) + O(B^{1-\delta}). \tag{2.9}$$

One obviously expects the leading coefficient of P in (2.9) to agree with Peyre's prediction, but there has so far been rather little investigation of the lower order terms. We now proceed to survey recent progress on the resolution of Conjectures 2.3 and 2.4 for del Pezzo surfaces of various degrees. Whenever we speak of these predictions we will always assume that the leading constant is that predicted by Peyre. All of the del Pezzo surfaces that we have discussed so far have been non-singular. In the following section we will also meet some singular ones.

2.3 Degree 3

Let us extend our definition of cubic del Pezzo surfaces slightly, by allowing a del Pezzo surface of degree 3 to be an irreducible cubic surface $S \subset \mathbb{P}^3$, which is not ruled by lines. This definition covers both singular and non-singular del Pezzo surfaces of degree 3. Given such a surface S defined over $\mathbb{Q}$, we may always find an absolutely irreducible cubic form $C \in \mathbb{Z}[x_1, x_2, x_3, x_4]$ such that S is defined by the equation $C = 0$.

2.3.1 Non-singular surfaces

We begin by considering the situation for non-singular cubic surfaces, for which one takes $U \subset S$ to be the open subset formed by deleting the 27 lines. It is very natural to enquire into the true nature of the error term in Manin's conjecture. This a question that Swinnerton-Dyer has addressed [116, Conjecture 2], inspired by comparisons with the explicit formulae from prime number theory. He has made the following bold reformulation of Conjecture 2.4.

Conjecture 2.5. *Suppose that $S \subset \mathbb{P}^3$ is a non-singular cubic surface, with $S(\mathbb{Q}) \neq \emptyset$. Then there exists a polynomial $P \in \mathbb{R}[x]$ of degree $\rho_S - 1$, a positive constant $\theta < \frac{1}{2}$ and a sequence of $\gamma_n \in \mathbb{C}$, such that for any $\varepsilon > 0$ we have*

$$N_U(B) = BP(\log B) + \Re e \sum_{n=1}^{\infty} \gamma_n B^{\frac{1}{2}+it_n} + O_\varepsilon(B^{\theta+\varepsilon}).$$

Here $t_n \in \mathbb{R}$ form a sequence of positive and monotonic increasing numbers, such that $\sum_{n=1}^{\infty} |\gamma_n|^2$ and $\sum_{n=1}^{\infty} t_n^{-1-\varepsilon}$ are both convergent for any $\varepsilon > 0$.

As in Conjecture 2.4, the leading coefficient of P should agree with Peyre's prediction. Conjecture 2.5 clearly implies that we should expect roughly square-root cancellation in the error term. We formalise this in the following conjecture.

Conjecture 2.6. *Suppose that $S \subset \mathbb{P}^3$ is a non-singular cubic surface, with $S(\mathbb{Q}) \neq \emptyset$. Then there exists a polynomial $P \in \mathbb{R}[x]$ of degree $\rho_S - 1$, such that for any $\varepsilon > 0$ we have*

$$N_U(B) = BP(\log B) + O_\varepsilon(B^{\frac{1}{2}+\varepsilon}).$$

The Manin conjectures lend themselves rather well to numerial testing and the setting of cubic surfaces is particularly fruitful. Peyre and Tschinkel [98, 99] have undertaken exentsive experiments for a number of diagonal cubic surfaces of the shape

$$a_1x_1^3 + a_2x_2^3 + a_3x_3^3 + a_4x_4^3 = 0, \tag{2.10}$$

for vectors $\mathbf{a} \in \mathbb{Z}^4$ with non-zero components. In Section 8.3 we will develop some of our own heuristics using the Hardy–Littlewood circle method, concentrating on the case $\mathbf{a} = (1, 1, 1, 1)$.

In order to provide evidence for Conjectures 2.5 and 2.6 we would like to check experimentally that

$$E_U(B) := \frac{N_U(B) - BP(\log B)}{\sqrt{B}} \tag{2.11}$$

has small order of magnitude for large B. This task is obviously complicated if $\rho_S > 1$ since then the polynomial P can have lower order coefficients that we don't yet have any predictions for. Thus we would like to work with a diagonal cubic surface $S \subset \mathbb{P}^3$ for which $\rho_S = 1$. The surface defined by

$$\mathbf{a} = (1, 1, 1, 2)$$

has this property and features in [69] and [98]. Although we will not present details here, one can calculate the numerical value of the predicted leading constant c_S to be

$$c_S = 2.07996061047\ldots,$$

which is probably correct to 10 decimal places. The author is grateful to Andrew Booker for help in performing this calculation. In Figure 2.1 we have plotted the behaviour of (2.11) for $50 \leqslant B \leqslant 3 \times 10^5$, with $P(\log B) = c_S$. A rather startling accordance with Conjecture 2.6 is readily observed: for height B up to $\frac{3}{10}$ of a million the maximum modulus of $E_U(B)$ is bounded by 10. The calculation of $N_U(B)$ for given B is carried out using the algorithm developed by Bernstein [5].

We have spent some time discussing what might possibly be dreamt of in terms of the counting function associated to a non-singular cubic surface. Let us

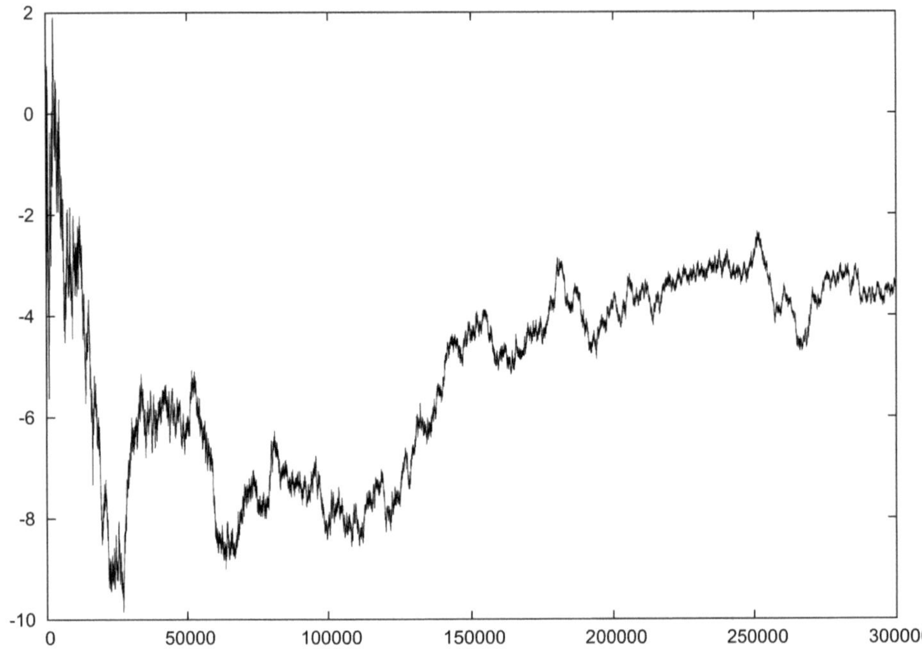

Figure 2.1: Behaviour of $E_U(B)$ for $50 \leqslant B \leqslant 3 \times 10^5$

now turn to a summary of what is actually known. The best upper bound available is

$$N_U(B) = O_{\varepsilon,S}(B^{\frac{4}{3}+\varepsilon}), \tag{2.12}$$

for any $\varepsilon > 0$. This is due to Heath-Brown [71] and applies to any non-singular cubic surface S that contains 3 coplanar lines defined over $\mathbb{Q}$. This covers in particular the Fermat cubic surface, obtained by taking $\mathbf{a} = (1, 1, 1, 1)$ in (2.10). Indeed, this surface contains the 3 lines

$$x_1 + x_2 = x_3 + x_4 = 0, \quad x_1 + x_3 = x_2 + x_4 = 0, \quad x_1 + x_4 = x_2 + x_3 = 0,$$

all of which lie in the plane $x_1 + x_2 + x_3 + x_4 = 0$.

In a subsequent investigation, Heath-Brown [73] returned to the subject of non-singular cubic surfaces, obtaining a conditional upper bound covering all such cubic surfaces. His work is subject to the following hypothesis concerning the growth rate of the rank of elliptic curves.

Conjecture 2.7. *Let E be an elliptic curve defined over $\mathbb{Q}$, with conductor C_E and rank r_E. Then we have $r_E = o(\log C_E)$.*

To put this into context, it is easily shown that $r_E = O(\log C_E)$ for any elliptic curve defined over $\mathbb{Q}$. Moreover, the theory of genera can be used to prove that Conjecture 2.7 holds when E has a rational point of order 2. A proof of these facts can be found in [73]. By work of Mestre [93], it also follows that the conjecture holds subject to the Birch–Swinnerton-Dyer conjecture together with the supposition that the associated Hasse–Weil L-function satisfies the Riemann hypothesis.

Assuming Conjecture 2.7, Heath-Brown [73] shows that (2.12) holds for all non-singular cubic surfaces. The key idea behind Heath-Brown's argument is to cover all of the rational points on S of height at most B with a small number of plane sections. This is basically Siegel's lemma, which we will meet in Chapter 7. One then needs to count points of bounded height on a large family of plane curves contained in the cubic surface. The generic plane section is a non-singular cubic curve, which is elliptic when it contains rational points. Conjecture 2.7 allows one to establish an excellent upper bound for the rational points of bounded height on an elliptic curve with only a very mild dependence on the particular curve.

The best unconditional upper bound for general non-singular cubic surfaces is due to Salberger. In work announced at the conference *Géométrie arithmétique et variétés rationnelles* in Luminy in 2007, he has shown that $N_U(B) = O_\varepsilon(B^{\frac{12}{7}+\varepsilon})$, for any $\varepsilon > 0$. The problem of proving lower bounds is somewhat easier. Under the assumption that S contains a pair of skew lines defined over $\mathbb{Q}$, Slater and Swinnerton-Dyer [114] have shown that $N_U(B) \gg_S B(\log B)^{\rho_S - 1}$, as predicted by the Manin conjecture. This does not cover the Fermat cubic surface, however, since the only skew lines contained in this surface are defined over $\mathbb{Q}(\sqrt{-3})$.

2.3.2 Singular surfaces

It turns out that much more can be said if one permits S to contain isolated singularities. For the remainder of this section let $S \subset \mathbb{P}^3$ be an irreducible cubic surface, which has only isolated singularities and is not a cone. The classification of singular cubic surfaces S is a well-established subject, and can be traced back to the work of Cayley [41] and Schläfli [108] over a century ago. A contemporary classification has since been given by Bruce and Wall [37], over $\overline{\mathbb{Q}}$. Around the time of Cayley and Schläfli there was something of an industry built up around producing plaster models of singular cubic surfaces. In Figure 2.2 we have included an image of a plaster model of the Cayley cubic surface defined below in (2.16). This is the unique cubic surface (over $\overline{\mathbb{Q}}$) with 4 singularities.

It follows from the investigations of Cayley, Schläfli, Bruce and Wall that S contains at most 4 isolated singularities and that these are all *rational double points*. As discussed by Coray and Tsfasman [47, Proposition 0.1], rational double points are singularities which can be resolved through a finite sequence of blow-ups of singular points, giving a *minimal desingularisation* $\pi : \widetilde{S} \rightarrow S$ of the surface, with $\widetilde{S}$ non-singular and $-K_{\widetilde{S}} = \pi^*(-K_S)$. The blow-ups produce (-2)-curves on $\widetilde{S}$. These exceptional divisors have *Dynkin diagrams* associated to them. For

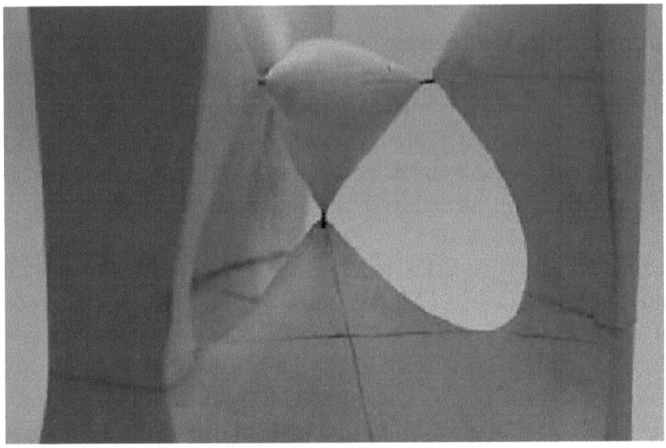

Figure 2.2: Plaster model of the Cayley cubic surface

arbitrary singular del Pezzo surfaces, including those of degree not equal to 3, the only possible Dynkin diagrams that arise in this way are

$$\mathbf{A}_n: \quad \alpha_1 \;\text{——}\; \alpha_2 \;\text{——}\; \cdots \;\text{——}\; \alpha_n$$

for $1 \leqslant n \leqslant 8$,

$$\mathbf{D}_n: \quad \begin{array}{ccccccc} & & \alpha_1 & & & & \\ & & | & & & & \\ \alpha_2 & \text{——} & \alpha_3 & \text{——} & \cdots & \text{——} & \alpha_n \end{array}$$

for $4 \leqslant n \leqslant 8$, and finally,

$$\mathbf{E}_n: \quad \begin{array}{ccccccccc} & & & & \alpha_1 & & & & \\ & & & & | & & & & \\ \alpha_2 & \text{——} & \alpha_2 & \text{——} & \alpha_3 & \text{——} & \cdots & \text{——} & \alpha_n \end{array}$$

for $6 \leqslant n \leqslant 8$. Here there is a line connecting two divisors in this diagram if and only if they meet in the minimal desingularisation. In general the intersection number of two distinct divisors represented in the Dynkin diagram is the number of edges between the corresponding vertices. The self-intersection number of a divisor in the diagram is -2.

It is customary to label the singularities according to the name of the corresponding Dynkin diagrams. For two singular cubic surfaces we say that they have the same singularity type if and only if the number and types of their singularities is the same. In Table 2.2 we have provided a classification of the 20 singularity types over $\overline{\mathbb{Q}}$ that arise in cubic surfaces, including the number of lines that each surface contains. Of course, if one is interested in a classification over the ground field $\mathbb{Q}$, then many more singularity types can occur (see Lipman [90], for example).

type	# lines	singularity
i	21	$\mathbf{A}_1$
ii	16	$2\mathbf{A}_1$
iii	15	$\mathbf{A}_2$
iv	12	$3\mathbf{A}_1$
v	11	$\mathbf{A}_1 + \mathbf{A}_2$
vi	10	$\mathbf{A}_3$
vii	9	$4\mathbf{A}_1$
viii	8	$2\mathbf{A}_1 + \mathbf{A}_2$
ix	7	$\mathbf{A}_1 + \mathbf{A}_3$
x	7	$2\mathbf{A}_2$
xi	6	$\mathbf{A}_4$
xii	6	$\mathbf{D}_4$
xiii	5	$2\mathbf{A}_1 + \mathbf{A}_3$
xiv	5	$\mathbf{A}_1 + 2\mathbf{A}_2$
xv	4	$\mathbf{A}_1 + \mathbf{A}_4$
xvi	3	$\mathbf{A}_5$
xvii	3	$\mathbf{D}_5$
xviii	3	$3\mathbf{A}_2$
ixx	2	$\mathbf{A}_1 + \mathbf{A}_5$
xx	1	$\mathbf{E}_6$

Table 2.2: Classification (over $\overline{\mathbb{Q}}$) of singular del Pezzo surfaces of degree 3

As an illustration of this consider the cubic surface

$$S_1 = \{x_1^2 x_3 + x_2 x_3^2 + x_4^3 = 0\}. \tag{2.13}$$

Up to isomorphism over $\overline{\mathbb{Q}}$ this is the unique cubic surface of type xx in the table. Its geometry is discussed at some length in [67]. The process of resolving the singularity gives 6 exceptional divisors $E_1, \ldots, E_6$ and produces the minimal desingularisation $\widetilde{S}_1$ of the surface S_1. If L denotes the strict transform of the unique line $x_3 = x_4 = 0$ on S_1, then $L, E_1, \ldots, E_6$ satisfy the intersection behaviour encoded in the Dynkin diagram

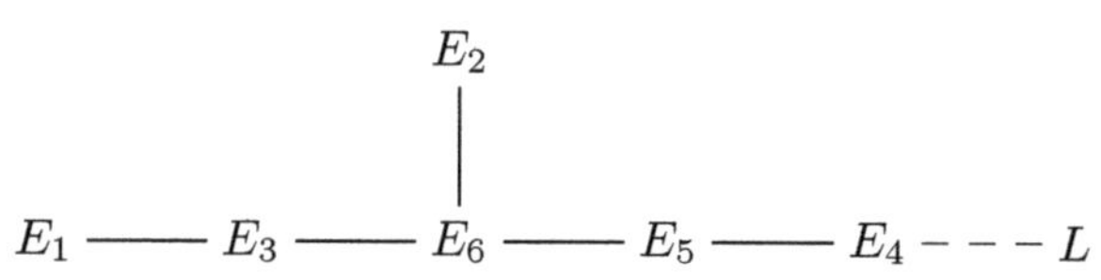

We have extended the $\mathbf{E}_6$ diagram by including the intersection behaviour of L with respect to the other divisors.

It turns out, as discussed in [37], that some types of surfaces do not have a single normal form, but an infinite family. This happens precisely for the surfaces of type `i`, `ii`, `iii`, `iv`, `v`, `vi` and `ix`. By [37, Lemma 4] the type `xii` surface, with a $\mathbf{D}_4$ singularity, is the only surface that has more than one normal form, but not a family. In fact it has precisely two normal forms, given by

$$S_2 = \{x_1x_2(x_1 + x_2) + x_4(x_1 + x_2 + x_3)^2 = 0\} \tag{2.14}$$

and

$$S_3 = \{x_1x_2x_3 + x_4(x_1 + x_2 + x_3)^2 = 0\}. \tag{2.15}$$

That these equations actually define distinct surfaces can be seen by calculating the corresponding Hessians in each case.

Having discussed the geometry of singular cubic surfaces to some extent, let us now turn to the quantitative arithmetic of these objects. The asymptotic formula (2.8) is still expected to hold, with ρ_S now taken to be the rank of the Picard group of $\widetilde{S}$. As usual $U \subset S$ is obtained by deleting all of the lines from S. When S is any surface from Table 2.2 in which all of its singularities and lines are defined over $\mathbb{Q}$, we say that the surface is *split*. In this case it follows that the Picard group of $\widetilde{S}$ has maximal rank 7 by (2.6), since $\mathrm{Pic}(\widetilde{S}) = \mathrm{Pic}_{\overline{\mathbb{Q}}}(\widetilde{S})$. For example, $[L], [E_1], \dots, [E_6]$ provide a basis for $\mathrm{Pic}(\widetilde{S}_1)$.

One would like to try and establish (2.8) for each surface in Table 2.2 with $\rho_S = 7$. Several del Pezzo surfaces are actually special cases of varieties for which the Manin conjecture is already known to hold. Recall that a variety of dimension D is said to be *toric* if it contains the algebraic group variety $\mathbb{G}_m^D$ as a dense open subset, whose natural action on itself extends to all of the variety. The Manin conjecture has been established for all toric varieties by Batyrev and Tschinkel [3]. It can be checked that the surface representing type `xviii` is toric, and is the only toric variety in the table. The surface has singularity type $3\mathbf{A}_2$ and is given by the equation

$$S_4 = \{x_1x_2x_3 + x_4^3 = 0\}.$$

Perhaps unsurprisingly, this particular surface has been studied by numerous authors, including la Bretèche [9], la Bretèche and Swinnerton-Dyer [18], Fouvry [59], Heath-Brown and Moroz [79] and Salberger [102]. Of the unconditional asymptotic formulae obtained, the most impressive is the first, which consists of an estimate like (2.9) for any $\delta \in (0, \frac{1}{8})$, with $\deg P = 6$. In [18] the analysis is pushed much further, with the outcome that under the Riemann hypothesis and the assumption that the zeros of the Riemann zeta function are all simple, one has

$$N_U(B) = BP(\log B) + \gamma B^{\frac{9}{11}} + \Re e \sum \gamma_n B^{\frac{13}{16} + i\frac{t_n}{8}} + O_\varepsilon(B^{\frac{4}{5}+\varepsilon})$$

for certain $\gamma, \gamma_n \in \mathbb{C}$ and any $\varepsilon > 0$. Here $U \subset S_4$ denotes the open subset formed by deleting the 3 lines from S_4 and $\frac{1}{2} + it_n$ runs through zeros of the

Riemann zeta function. This constitutes the most compelling evidence that we have for Conjecture 2.5, which we recall was phrased in terms of non-singular cubic surfaces there. One actually has $\gamma \in \mathbb{R} \setminus \{0\}$ in the above and it would be interesting to determine whether or not the exponents $\frac{9}{11}$ or $\frac{13}{16}$ are somehow related to the geometry of the surface.

Given a linear algebraic group G, a projective variety V is said to be an *equivariant compactification* of G if there exists a G-action $G \times V \to V$, with $(g, v) \mapsto g \circ v$, together with a Zariski dense open subset $U \subseteq V$ and an isomorphism $\phi : U \to G$ such that $\phi(g \circ v) = g\phi(v)$ for all $g \in G$ and $v \in U$. Recall that the map $G \times V \to V$ is a G-action if $(gh) \circ v = g \circ (h \circ v)$ and $e_G \circ v = v$ for all $g, h \in G$ and $v \in V$. One can check that a toric surface is an equivariant compactification of $\mathbb{G}_m^2$. It is natural to ask whether any other surfaces in Table 2.2 arise as equivariant compactifications of linear algebraic groups. In private communication with the author Ulrich Derenthal and Daniel Loughran have shown that no cubic surfaces arise as equivariant compactifications of $\mathbb{G}_a^2$.

The next surface to have received serious attention is the *Cayley cubic surface*

$$S_5 = \{x_1x_2x_3 + x_1x_2x_4 + x_1x_3x_4 + x_2x_3x_4 = 0\}, \tag{2.16}$$

which is the type `vii` surface in the table. Heath-Brown [76] has shown that there exist absolute constants $A_1, A_2 > 0$ such that

$$A_1 B(\log B)^6 \leqslant N_U(B) \leqslant A_2 B(\log B)^6.$$

An estimate of precisely the same form has been obtained by Browning [22] for the $\mathbf{D}_4$ surface S_3 in (2.15). In both cases the lines and singularities in the surface are all defined over $\mathbb{Q}$, so that the surfaces are split. Thus the corresponding Picard groups have rank 7 and the exponents of B and $\log B$ agree with Manin's prediction. In Chapter 6 we will establish an upper bound for the remaining $\mathbf{D}_4$ cubic surface S_2 in (2.14). This will involve two basic attacks. First we will give a comprehensive account of the upper bound $N_U(B) = O_\varepsilon(B^{1+\varepsilon})$, for any $\varepsilon > 0$. Next, by making use of the work in [22], we will establish the following finer result.

Theorem 2.1. *Let S_2 be given by (2.14). We have $N_U(B) \ll B(\log B)^6$.*

The cubic surface S_2 contains the unique singular point $[0, 0, 0, 1]$, together with 6 lines defined over $\mathbb{Q}$. Thus the surface is split and it follows that Theorem 2.1 agrees with the Manin conjecture.

A further surface to have been studied extensively is the $\mathbf{E}_6$ cubic surface S_1 that we discussed above. The particular geometry of this surface was investigated by Hassett and Tschinkel [67]. Building on this, la Bretèche, Browning and Derenthal [19] have succeeded in establishing the Manin conjecture for S_1. In fact an asymptotic formula of the shape (2.9) is achieved, with P of degree 6 and any $\delta \in (0, \frac{1}{11})$. It should be remarked that Joyce [85] has independently established the Manin conjecture for S_1 in his doctoral thesis, albeit with a weaker error term of $O(B(\log B)^5)$.

Figure 2.3: Points of height $\leqslant 100$ on the $\mathbf{D}_5$ cubic surface.

The final singular cubic surface to have been examined is the $\mathbf{D}_5$ cubic surface

$$S_6 = \{x_1x_3^2 + x_1^2x_4 + x_2^2x_3 = 0\},$$

representing type `xvii` in Table 2.2. Note that S_6 contains the unique singularity $[0,0,0,1]$ of type $\mathbf{D}_5$ and three lines, each of which is defined over $\mathbb{Q}$. In Figure 2.3, which was constructed by Derenthal, an illustration is given of all the rational points of height $\leqslant 100$ on this surface. The outcome of the work of Browning and Derenthal [27] is an asymptotic formula of the shape

$$N_U(B) = c_{S_6}B(\log B)^6 + O\big(B(\log B)^5(\log\log B)\big),$$

where the leading constant c_{S_6} is that predicted by Peyre. This therefore confirms (2.8) in this particular case.

The treatment of all of the singular cubic surfaces mentioned in this section rely on first parametrising the rational points on the surface in a suitable way. This "passage to the universal torsor" will be discussed briefly in Section 2.6. We will get a flavour of some of the techniques involved in Chapter 5.

2.4 Degree 4

A quartic del Pezzo surface $S \subset \mathbb{P}^4$ that is defined over $\mathbb{Q}$ can be given as the zero locus of a pair of quadratic forms $Q_1, Q_2 \in \mathbb{Z}[x_1, \ldots, x_5]$. Again we do not stipulate that the surface should be non-singular, but do insist that S is not a cone. As usual let $U \subset S$ denote the open subset formed by deleting all of the lines from S. Our aim in this section is to discuss Conjectures 2.3 and 2.4 in the context of quartic del Pezzo surfaces.

2.4.1 Non-singular surfaces

Let us begin with the situation for non-singular surfaces $S \subset \mathbb{P}^4$, which arise when the Jacobian matrix $(\nabla Q_1, \nabla Q_2)$ has full rank throughout S. In this setting there are 16 lines to delete in forming U. Analogously to Section 2.1.4, where the 27 lines on a non-singular cubic surface were discussed, it is easy to give a description of the 16 lines following Manin [91]. Non-singular quartic del Pezzo surfaces are obtained by blowing up $\mathbb{P}^2$ along a collection of 5 points $P_1, \ldots, P_5$, no 3 of which are collinear. The 16 lines arise in the following way:

- there are 5 (-1)-curves E_i above P_i, for $1 \leqslant i \leqslant 5$;
- there are 10 strict transforms $L_{i,j}$ of the lines going through precisely 2 points P_i, P_j, for $1 \leqslant i < j \leqslant 5$;
- there is 1 strict transform Q of the conic going through all 5 points.

If Λ is the strict transform of a line in $\mathbb{P}^2$ that does not go through any of the P_i, then a basis of the geometric Picard group $\mathrm{Pic}_{\overline{\mathbb{Q}}}(S)$ is given by $[\Lambda], [E_1], \ldots, [E_5]$.

One of the best results available is due to Salberger. In work communicated at the conference *Higher-dimensional varieties and rational points* in Budapest in 2001, he establishes the estimate

$$N_U(B) = O_{\varepsilon,S}(B^{1+\varepsilon}) \tag{2.17}$$

for any $\varepsilon > 0$, provided that the surface contains a conic defined over $\mathbb{Q}$. The latter condition forces S to have a conic bundle structure, as the following result will demonstrate.

Lemma 2.2. *Let $S \subset \mathbb{P}^4$ be a non-singular del Pezzo surface of degree 4 with $S(\mathbb{Q}) \neq \emptyset$ and suppose that S contains a conic defined over $\mathbb{Q}$. Then S takes the shape*

$$\begin{cases} Q_1(x_1, \ldots, x_5) := x_1x_2 - x_3x_4 = 0, \\ Q_2(x_1, \ldots, x_5) = 0, \end{cases} \tag{2.18}$$

for a certain quadratic form $Q_2 \in \mathbb{Z}[x_1, \ldots, x_5]$.

Proof. Suppose S is defined by quadratic forms $Q_1, Q_2 \in \mathbb{Z}[x_1, \dots, x_5]$. Our task is to find a non-singular linear change of variables that takes Q_1 into the shape given in (2.18).

By assumption S contains a conic defined over $\mathbb{Q}$. We may assume without loss of generality that the conic is contained in the plane $x_4 = x_5 = 0$. This means that there exists a ternary quadratic form $Q \in \mathbb{Z}[x_1, x_2, x_3]$ such that $Q \mid Q_i(x_1, x_2, x_3, 0, 0)$ for $i = 1, 2$, whence $Q_i(x_1, x_2, x_3, 0, 0) = \mu_i Q$ for certain $\mu_1, \mu_2 \in \mathbb{Z}$. We may therefore assume that S is defined by the pair of quadratic forms

$$Q_i(\mathbf{x}) = \mu_i Q(x_1, x_2, x_3) + L_i(x_1, x_2, x_3)x_4 + M_i(x_1, x_2, x_3)x_5 + P_i(x_4, x_5),$$

where $L_i, M_i \in \mathbb{Q}[x_1, x_2, x_3]$ are linear and $P_i \in \mathbb{Q}[x_4, x_5]$ is quadratic. Clearly one may eliminate the appearance of $Q(x_1, x_2, x_3)$ from one of the equations. Proceeding under the assumption that $\mu_1 = 0$, as we clearly may, we conclude that there exist linear forms $L, M \in \mathbb{Z}[x_1, x_2, x_3]$ and an integral binary quadratic form P such that

$$Q_1(\mathbf{x}) = L(x_1, x_2, x_3)x_4 + M(x_1, x_2, x_3)x_5 + P(x_4, x_5).$$

An examination of the Jacobian criterion is enough to ensure that neither L nor M vanishes when S is non-singular. Suppose that $P(x, y) = ax^2 + bxy + cy^2$, for integers a, b, c. Then on carrying out the non-singular change of variables

$$y_2 = L(x_1, x_2, x_3) + ax_4 + bx_5, \quad y_3 = -M(x_1, x_2, x_3) + cx_5, \quad y_i = x_i,$$

for $i = 1, 4, 5$, we find that Q_1 is taken into the form $y_2y_4 - y_3y_5$, as required to complete the proof of the lemma. □

Let $S \subset \mathbb{P}^4$ be as in the statement of Lemma 2.2. One may then define a pair of conic bundle morphisms $f_i : S \to \mathbb{P}^1$, for $i = 1, 2$. Specifically, for any $x \in S$, one takes

$$f_1(x) = \begin{cases} [x_1, x_3], & \text{if } (x_1, x_3) \neq (0, 0), \\ [x_2, x_4], & \text{otherwise,} \end{cases}$$

and

$$f_2(x) = \begin{cases} [x_1, x_4], & \text{if } (x_1, x_4) \neq (0, 0), \\ [x_2, x_3]; & \text{otherwise.} \end{cases}$$

For a given point $x \in S(\mathbb{Q})$ of height $H(x) = H_4(x) \leqslant B$, it follows from the general theory of height functions (see Serre [111, Chapter 2], for example) that

$$H_4(x) \asymp H_1(f_1(x))H_1(f_2(x)),$$

where we have written H_{n-1} for the usual exponential height on $\mathbb{P}^{n-1}(\mathbb{Q})$. Hence there exists an index i such that $x \in f_i^{-1}(t)$ for some $t \in \mathbb{P}^1(\mathbb{Q})$ of height $O(\sqrt{B})$.

The idea is now to count rational points of bounded height on the fibres $f_1^{-1}(t)$ and $f_2^{-1}(t)$, uniformly for points $t \in \mathbb{P}^1(\mathbb{Q})$ of height $O(\sqrt{B})$. This is the strategy adopted by Salberger in his proof of (2.17), but the idea can be traced back to Heath-Brown's [71] proof of (2.12). In fact the reader may wish to entertain comparisons with the "hyperbola method" devised by Dirichlet in his treatment of the divisor problem.

Leung [89] has revisited Salberger's proof of (2.17), ultimately replacing the B^ε by a small power of $\log B$ under the same hypotheses. It sometimes happens in analytic number theory that a method of attack that yields a good upper bound can be pushed into giving the full asymptotic formula if one is prepared to work hard enough. This is the case here. In recent work of la Bretèche and Browning [15], Conjecture 2.3 has been resolved for a concrete non-singular quartic del Pezzo surface for the first time. Specifically, the following result has been established.

Theorem 2.2. *Let $S \subset \mathbb{P}^4$ be given by (2.18), with*

$$Q_2(x_1, \dots, x_5) := x_1^2 + x_2^2 + x_3^2 - x_4^2 - 2x_5^2.$$

Then we have

$$N_U(B) = c_S B(\log B)^{\rho_S - 1}\Big(1 + O\Big(\frac{1}{\log\log B}\Big)\Big),$$

where $\rho_S = 5$ is the rank of the Picard group.

Note that S contains lines that are defined over $\mathbb{Q}(i)$ and so S is not split. The first stage in the proof of Theorem 2.2 involves a reduction of the problem to one involving counting points on the conic fibres. This ultimately leads one to count $(a, b, x, y, z) \in \mathbb{N}^5$ such that

$$C_{a,b}: \quad (a^2 - b^2)x^2 + (a^2 + b^2)y^2 = 2z^2, \tag{2.19}$$

with

$$\gcd(a, b) = \gcd(x, y) = 1, \quad ab, xy \neq 1, \quad \max\{a, b\}\max\{x, y\} \leqslant B.$$

By symmetry one can restrict to the case in which $\max\{a, b\} \leqslant \max\{x, y\}$, so that really one is interested in a, b such that $\max\{a, b\} \leqslant \sqrt{B}$. This corresponds to the argument involving height machinery that was discussed above in the context of Salberger's proof of (2.17). For fixed coprime a, b such that $ab \neq 1$ it is clear that $C_{a,b} \subset \mathbb{P}^2$ defines a non-singular plane conic with discriminant equal to $-2(a^4 - b^4)$. In Figure 2.4 we have sketched the affine part of the system of fibres $\{C_{x,1}\}_{x \in \mathbb{A}^1}$.

The next stage in the argument hinges upon the observation that for each $a, b \in \mathbb{N}$ the conic $C_{a,b}$ always contains the rational point

$$\xi = [1, -1, a],$$

corresponding to the fact that the morphism $S \to \mathbb{P}^1$ has a section. In the classical manner we can use this point to parametrise $C_{a,b}(\mathbb{Q})$, by considering the residual

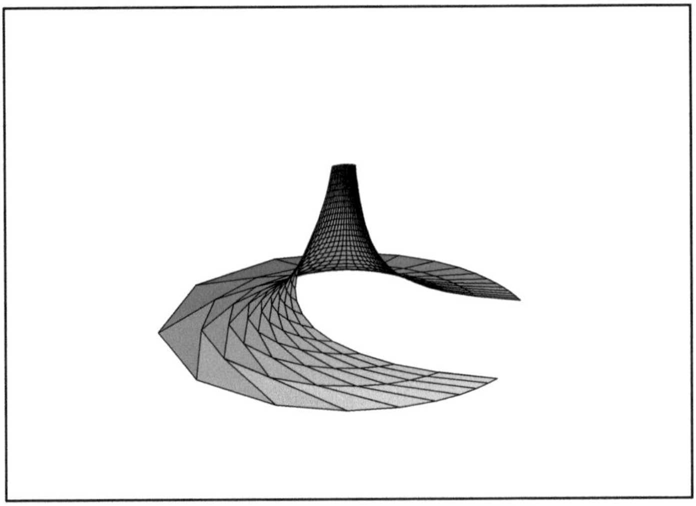

Figure 2.4: The fibres $\{C_{x,1}\}_{x\in\mathbb{A}^1}$

intersection with $C_{a,b}$ of an arbitrary line through ξ. One finds that the general point on the conic (2.19) is given by taking $(x,y,z) = (Q_1(s,t), Q_2(s,t), Q_3(s,t))$, with

$$\begin{aligned} Q_1(s,t) &= -2s^2 - (a^2-b^2)t^2 + 4ast,\\ Q_2(s,t) &= 2s^2 - (a^2-b^2)t^2,\\ Q_3(s,t) &= 2as^2 - 2(a^2-b^2)st + a(a^2-b^2)t^2. \end{aligned}$$

Ultimately one is led to counting primitive lattice points (s,t) in the plane which are restricted to lie in a lopsided region with piecewise continuous boundary. One needs to estimate this quantity with complete uniformity with respect to the parameters a and b. This is the most challenging part of the argument and we refer the reader to [15] for details. In the final stages of the paper one is led to analyse the divisor sums $D(X;\varphi,F)$ defined in (1.13). Here one is interested in the special case $F(x_1,x_2) = x_1^4 - x_2^4$, and φ is taken to be an arithmetic function which is close to the divisor function τ, in the sense that one has a Dirichlet convolution $\phi = \tau * h$ for a "small" arithmetic function h. A treatment of this problem is corralled into an allied investigation [16].

Perhaps unsurprisingly, it turns out that Heath-Brown's conditional treatment [73] of non-singular cubic surfaces can be adapted to handle arbitrary degree 4 surfaces. We will establish the following result in Chapter 7.

Theorem 2.3. *Assume that Conjecture 2.7 holds. Let $S \subset \mathbb{P}^4$ be a non-singular del Pezzo surface of degree 4. Then we have $N_U(B) = O_{\varepsilon,S}(B^{\frac{5}{4}+\varepsilon})$.*

We have spent some time discussing the situation for conic bundle quartic del Pezzo surfaces. The best unconditional upper bound for general non-singular quartic del Pezzo surfaces follows from Salberger's investigation [106], who has shown that $N_U(B) = O_{\varepsilon,S}(B^{\frac{3}{2}+\varepsilon})$, for any $\varepsilon > 0$.

2.4.2 Singular surfaces

As with cubic surfaces, much more can be said if one permits S to contain isolated singularities. For the remainder of this section let $S \subset \mathbb{P}^4$ be an irreducible intersection of two quadric hypersurfaces, which has only isolated singularities and is not a cone. Let $\widetilde{S}$ be the minimal desingularisation of S. Then the asymptotic formula (2.8) is still expected to hold, with ρ_S now taken to be the rank of the Picard group of $\widetilde{S}$ and $U \subset S$ obtained by deleting all of the lines from S. In particular, when S is split, one always has $\rho_S = 6$.

The classification of singular quartic del Pezzo surfaces can be extracted from the work of Hodge and Pedoe [80, Book IV, Section XIII.11], where it is phrased in terms of the so-called Segre symbol. The *Segre symbol* of a matrix $\mathbf{M} \in M_5(\mathbb{C})$ is defined as follows. If the Jordan form of $\mathbf{M}$ has Jordan blocks of sizes $a_1, \dots, a_n$, with $a_1 + \cdots + a_n = 5$, then the Segre symbol is the symbol

$$(a_1, \dots, a_n)$$

with extra parentheses around the Jordan blocks with equal eigenvalues. Suppose that our quartic del Pezzo surface S is defined by a pair of quadric hypersurfaces, with underlying symmetric matrices $\mathbf{A}, \mathbf{B} \in M_5(\mathbb{Q})$. Then the Segre symbol of S is defined to be the Segre symbol associated to $\mathbf{A}^{-1}\mathbf{B}$. A crucial property of the Segre symbol is that it does not depend on the choice of $\mathbf{A}$ and $\mathbf{B}$ in the pencil of quadrics defining S. Since we are assuming that S is not a cone, one may always suppose that $\mathbf{A}, \mathbf{B}$ are chosen so that $\mathbf{A}$ has full rank.

To illustrate the calculation of the Segre symbol, let us consider the surface S defined by the pair of equations

$$\begin{cases} x_1x_2 + x_3x_4 = 0, \\ x_1x_4 + x_2x_3 + x_3x_5 + x_4x_5 = 0. \end{cases} \tag{2.20}$$

Let $\mathbf{A}, \mathbf{B} \in M_5(\mathbb{Q})$ denote the underlying matrices of the first and second equations, respectively. Then $\mathbf{A}$ has rank 4, and so we replace it with $\mathbf{A}+2\mathbf{B}$, which has full rank. A simple calculation reveals that the matrix $(\mathbf{A} + 2\mathbf{B})^{-1}\mathbf{B}$ has Jordan form

$$\mathbf{J} = \begin{pmatrix} 0 & 0 & 0 & 0 & 0 \\ 0 & 1 & 0 & 0 & 0 \\ 0 & 0 & \frac{1}{3} & 0 & 0 \\ 0 & 0 & 0 & \frac{1}{2} & 1 \\ 0 & 0 & 0 & 0 & \frac{1}{2} \end{pmatrix}.$$

This matrix has 4 Jordan blocks, one of size 2 and the rest of size 1. The eigenvalues associated to the different Jordan blocks are all different, and so it follows that the surface (2.20) has Segre symbol $(2,1,1,1)$.

So far we have given a very easy way to check the isomorphism type of a given singular del Pezzo surface of degree 4. How do we match this up with a classification according to the singularity type, as in our discussion of cubic surfaces in Table 2.2? It turns out that up to isomorphism over $\overline{\mathbb{Q}}$, there are 15 possible singularity types for S. Over $\overline{\mathbb{Q}}$, Coray and Tsfasman [47, Proposition 6.1] have calculated the extended Dynkin diagrams for all of the 15 types, and Knörrer [86] has determined the precise correspondence between the singularity type and the Segre symbol. Table 2.3 is extracted from this body of work, and matches each possible singularity type with the Segre symbol, and the number of lines that the surface contains.

type	Segre symbol	# lines	singularity
i	(2,1,1,1)	12	$\mathbf{A}_1$
ii	(2,2,1)	9	$2\mathbf{A}_1$
iii	((1,1),1,1,1)	8	$2\mathbf{A}_1$
iv	(3,1,1)	8	$\mathbf{A}_2$
v	((1,1),2,1)	6	$3\mathbf{A}_1$
vi	(3,2)	6	$\mathbf{A}_1+\mathbf{A}_2$
vii	(4,1)	5	$\mathbf{A}_3$
viii	((2,1),1,1)	4	$\mathbf{A}_3$
ix	((1,1),(1,1),1)	4	$4\mathbf{A}_1$
x	((1,1),3)	4	$2\mathbf{A}_1+\mathbf{A}_2$
xi	((2,1),2)	3	$\mathbf{A}_1+\mathbf{A}_3$
xii	(5)	3	$\mathbf{A}_4$
xiii	((3,1),1)	2	$\mathbf{D}_4$
xiv	((2,1),(1,1))	2	$2\mathbf{A}_1+\mathbf{A}_3$
xv	((4,1))	1	$\mathbf{D}_5$

Table 2.3: Classification (over $\overline{\mathbb{Q}}$) of singular del Pezzo surfaces of degree 4

In general, given a particular Segre symbol, it is not trivial to determine explicit equations that define a singular del Pezzo surface of degree 4 having this symbol. Nonetheless in Table 2.4 we have done precisely this for each Segre symbol that occurs. In doing so we have retrieved some of the calculations carried out by Derenthal [53]. An important feature of the table is that the surfaces recorded are split over $\mathbb{Q}$. It remains a significant open challenge to establish the Manin conjecture for the 15 surfaces given in Table 2.4. This will undoubtedly prove a key stepping stone on the way towards a resolution of the conjecture for all del Pezzo surfaces. There is huge potential for further work in this area, and one of the themes of this book is to show that analytic number theorists are well placed to make a contribution.

type	$Q_1(\mathbf{x})$	$Q_2(\mathbf{x})$
i	$x_1x_2 - x_3x_4$	$x_1x_4 - x_2x_3 + x_3x_5 + x_4x_5$
ii	$x_1x_2 - x_3x_4$	$x_1x_4 - x_2x_3 + x_3x_5 + x_5^2$
iii	$x_1x_2 - x_3^2$	$x_1x_3 - x_2x_3 + x_4x_5$
iv	$x_1x_2 - x_3x_4$	$(x_1 + x_2 + x_3 + x_4)x_5 - x_3x_4$
v	$x_1x_2 - x_3^2$	$x_2x_3 + x_3^2 + x_4x_5$
vi	$x_1x_2 - x_3x_4$	$x_1x_5 + x_2x_3 + x_4x_5$
vii	$x_1x_2 - x_3x_4$	$x_1x_4 + x_2x_4 + x_3x_5$
viii	$x_1x_4 - (x_2 - x_3)x_5$	$(x_1 + x_4)(x_2 + x_3) + x_2x_3$
ix	$x_1x_2 - x_3^2$	$x_3^2 - x_4x_5$
x	$x_1x_2 - x_3^2$	$x_2x_3 - x_4x_5$
xi	$x_1x_4 - x_3x_5$	$x_1x_2 + x_2x_4 + x_3^2$
xii	$x_1x_2 - x_3x_4$	$x_1x_5 + x_2x_3 + x_4^2$
xiii	$x_1x_4 - x_2x_5$	$x_1x_2 + x_2x_4 + x_3^2$
xiv	$x_1x_2 - x_3^2$	$x_1^2 - x_4x_5$
xv	$x_1x_2 - x_3^2$	$x_1x_5 + x_2x_3 + x_4^2$

Table 2.4: Split surfaces representing the 15 singularity types

Whereas they share the same singularity type, the surfaces of type `vii` and `viii` differ because in the former there are 5 lines, 4 of which pass through the singularity, whereas in the latter all 4 lines pass through the singularity. Similarly, an important difference between the surfaces of type `ii` and `iii` in Tables 2.3 and 2.4 is that for the surface of type `ii`, the line joining the two singularities is contained in the surface, whereas for the surface of type `iii` it is not.

As usual, let $\widetilde{S}$ denote the minimal desingularisation of any surface S from Table 2.4. Then the Picard group of $\widetilde{S}$ has rank $\rho_S = 6$. The goal is then to try and establish (2.8) for each S. As in the case of singular cubic surfaces several of the surfaces are actually special cases of varieties for which the Manin conjecture is already known to hold. Thus it can be shown that the surfaces representing types `ix`, `x`, `xiv` are all toric, so that (2.8) already holds in these cases by the work of Batyrev and Tschinkel [3]. As illustrated in Exercise 2.5, there is a very real sense in which these surfaces are the easiest to deal with in our list.

It has also been shown by Chambert-Loir and Tschinkel [42] that the Manin conjecture is true for equivariant compactifications of the algebraic group $\mathbb{G}_a^2$. Although identifying such surfaces in the table is not entirely routine, it transpires that the type `xv` surface (with a $\mathbf{D}_5$ singularity) is covered by this work. One can show that it is the only surface on the list arising in this way. La Bretèche and Browning [12] have provided an independent proof of the Manin conjecture for this particular surface. In addition to obtaining a finer asymptotic formula of the shape given in (2.9), this work has provided a fruitful line of attack for several other singular del Pezzo surfaces.

In Table 2.5 we have recorded a list of progress towards the final resolution of the Manin conjecture for the split singular del Pezzo surfaces of degree 4. We have included the relevant references to the literature, but we will not pay attention here to the quality of the error term in the asymptotic formula. There remains plenty left to do! For the $3\mathbf{A}_1$ surface of type `v` an upper bound of the correct order of magnitude has been provided by Browning [23].

type	singularity	type of estimate achieved
`ix`	$4\mathbf{A}_1$	asymptotic formula [3]
`x`	$2\mathbf{A}_1 + \mathbf{A}_2$	asymptotic formula [3]
`xi`	$\mathbf{A}_1 + \mathbf{A}_3$	asymptotic formula [54]
`xii`	$\mathbf{A}_4$	asymptotic formula [26]
`xiii`	$\mathbf{D}_4$	asymptotic formula [55]
`xiv`	$2\mathbf{A}_1 + \mathbf{A}_3$	asymptotic formula [3]
`xv`	$\mathbf{D}_5$	asymptotic formula [12]

Table 2.5: Progress for the split singular del Pezzo surfaces of degree 4

It is also interesting to try and establish the Manin conjecture for singular del Pezzo surfaces of degree 4 that are not split over the ground field. In further work of la Bretèche and Browning [13], the Manin conjecture is established for the surface

$$\begin{cases} x_1x_2 - x_3^2 = 0, \\ x_1^2 + x_2x_5 + x_4^2 = 0. \end{cases}$$

This surface has a $\mathbf{D}_4$ singularity and is isomorphic over $\mathbb{Q}(i)$ to the surface of type `xiii` in Table 2.4. The Picard group of $\widetilde{S}$ has rank 4 in this case, and an asymptotic formula of the shape (2.9) is obtained for any $\delta \in (0, \frac{3}{32})$, with P a polynomial of degree 3.

When the 2 singular points are defined over a quadratic extension of $\mathbb{Q}$, the surface of type `iii` in Table 2.3 is called an *Iskovskikh surface*. There is ample evidence available (see Coray and Tsfasman [47], for example) to the effect that Iskovskikh surfaces are the most arithmetically interesting surfaces among the singular del Pezzo surfaces of degree 4. Indeed these are the only such surfaces for which the Hasse principle or weak approximation can fail. Note that a projective variety V is said to satisfy the *Hasse principle* if $V(\mathbb{Q}) \neq \emptyset$ if and only if $V(\mathbb{R}) \neq \emptyset$ and $V(\mathbb{Q}_p) \neq \emptyset$ for each prime p. It is said to satisfy *weak approximation* if the map

$$V(\mathbb{Q}) \to V(\mathbb{R}) \times \prod_p V(\mathbb{Q}_p)$$

has dense image under the product topology.

The general shape taken by an Iskovskikh surface is

$$S: \quad \begin{cases} x_1x_2 - x_3^2 = 0, \\ Q(x_1,x_2,x_3) - x_4^2 - ax_5^2 = 0, \end{cases}$$

where $a \in \mathbb{Z}$ is not equal to a square and $Q \in \mathbb{Z}[x_1,x_2,x_3]$ is a quadratic form. One easily checks that S has eight lines and two conjugate singular points

$$e_1 = [0,0,0,1,\sqrt{a}], \quad e_2 = [0,0,0,1,-\sqrt{a}].$$

The line connecting e_1 and e_2 is given by $x_1 = x_2 = x_3 = 0$, which clearly does not lie in S, whence it follows that S is an Iskovskikh surface. Now it follows from work of Browning [25] that

$$N_S(B) = O(B(\log B)^{\rho_S - 1}),$$

where ρ_S is the rank of the Picard group of the associated minimal desingularisation $\widetilde{S}$.

A much deeper result has been attained by la Bretèche, Browning and Peyre [20], wherein the Manin conjecture is established for the family of surfaces arising when the binary form $Q(u^2,uv,v^2)$ has non-zero discriminant and factorises as a product of four linear forms in u,v over $\mathbb{Q}$. It turns out that for an infinite family of such surfaces the map $S(\mathbb{Q}) \to S(\mathbb{R}) \times S(\mathbb{Q}_2)$ has non-dense image, whence there is a failure of weak approximation. This failure is explained by the Brauer-Manin obstruction, which in the light of work due to Colliot-Thélène, Sansuc and Swinnerton-Dyer [45, 46] is the only obstruction to the Hasse principle and weak approximation for Iskovskikh surfaces. A key ingredient in [20] is an asymptotic formula for the divisor sum $D(X;r,F)$ given in (1.13), where if χ is the real non-principle character modulo 4, then

$$r(n) := 4\sum_{d|n}\chi(d)$$

counts the number of representations of a positive integer n as a sum of 2 squares. Furthermore, one needs to understand the case in which F factorises completely as a product of pairwise coprime linear forms defined over $\mathbb{Z}$. The estimation of this sum is the object of parallel work due to la Bretèche and Browning [14], which extends earlier work of Heath-Brown [75].

2.5 Degree $\geqslant 5$

It turns out that all del Pezzo surfaces of degree $d \geqslant 7$ are toric [53, Proposition 8]. Moreover, all non-singular del Pezzo surfaces of degree $d \geqslant 6$ are toric. Thus (2.8) already holds in these cases by the work of Batyrev and Tschinkel [3]. For non-singular del Pezzo surfaces $S \subset \mathbb{P}^5$ of degree 5, the situation is rather less satisfactory. In fact there are very few instances for which the Manin conjecture has been established. The most significant of these is due to la Bretèche [10], who has proved the conjecture for the split non-singular del Pezzo surface S of degree 5, in which the 10 lines are all defined over $\mathbb{Q}$. To be precise, if $U \subset S$ denotes the open subset formed by deleting the lines from S, then la Bretèche shows that $N_U(B) \sim cB(\log B)^4$ for a certain constant $c > 0$. This confirms Conjecture 2.3, since we have seen in (2.6) that $\mathrm{Pic}(S) \cong \mathbb{Z}^5$. The other major achievement in the setting of quintic del Pezzo surfaces is a result of la Bretèche and Fouvry [17], where the Manin conjecture is established for a surface that is not split, but contains lines defined over $\mathbb{Q}(i)$.

type	# lines	singularity
i	4	$\mathbf{A}_1$
ii	3	$\mathbf{A}_1$
iii	2	$2\mathbf{A}_1$
iv	2	$\mathbf{A}_2$
v	1	$\mathbf{A}_1 + \mathbf{A}_2$

Table 2.6: Classification (over $\overline{\mathbb{Q}}$) of singular del Pezzo surfaces of degree 6

So far we have only discussed the situation for non-singular del Pezzo surfaces of degree $d \geqslant 5$. Let us now turn to the singular setting. When $d = 6$, it emerges that there exist surfaces that are not toric, and so are not covered by [3]. We will focus attention on the situation for del Pezzo surfaces of degree 6, following the geometric investigation of Derenthal [53]. In view of [47, Proposition 8.3], Table 2.6 lists all of the possible singularity types of degree 6 singular del Pezzo surfaces. The surfaces of type `i`, `iii` and `v` are all toric and so do not interest us here. It turns out that the remaining two surfaces, of type `ii` and `iv`, can be shown to arise as equivariant compactifications of $\mathbb{G}_a^2$. Thus these examples are covered by the work of Chambert-Loir and Tschinkel [42], whence the Manin conjecture is known to hold for all del Pezzo surfaces of degree 6.

In order to illustrate some of the techniques involved, however, we will give a self-contained proof of Conjecture 2.3 for the split surface of type `ii` in Table 2.6. Any singular del Pezzo surface of degree 6 can be realised as the intersection of 9 quadrics in $\mathbb{P}^6$. The type `ii` surface is cut out by the system of equations

$$\begin{aligned} x_1^2 - x_2x_4 &= x_1x_5 - x_3x_4 = x_1x_3 - x_2x_5 = x_1x_6 - x_3x_5 \\ &= x_2x_6 - x_3^2 = x_4x_6 - x_5^2 = x_1^2 - x_1x_4 + x_5x_7 \\ &= x_1^2 - x_1x_2 - x_3x_7 = x_1x_3 - x_1x_5 + x_6x_7 = 0. \end{aligned} \tag{2.21}$$

Let $\widetilde{S}$ denote the minimal desingularisation of S. We will establish the following result in Chapter 5.

Theorem 2.4. *Let $S \subset \mathbb{P}^6$ be the $\mathbf{A}_1$ surface given by (2.21). Then there exist constants $c_1, c_2 \geqslant 0$ such that*

$$N_U(B) = c_1 B(\log B)^3 + c_2 B(\log B)^2 + O\big(B \log B\big),$$

where

$$c_1 = \frac{\sigma_\infty}{144} \prod_p \Big(1 - \frac{1}{p}\Big)^4 \Big(1 + \frac{4}{p} + \frac{1}{p^2}\Big)$$

and

$$\sigma_\infty = 6 \int_{\{u,t,v \in \mathbb{R}:\ 0<u,ut^2,uv^2,|tv(t-v)|\leqslant 1\}} dt du dv. \tag{2.22}$$

It follows from (2.6) that $\mathrm{Pic}(\widetilde{S}) \cong \mathbb{Z}^4$ since S is split. Thus the exponents of B and $\log B$ are in complete agreement with Conjecture 2.3. Although we will not give details here, it can be shown that the value of the constant c_1 also confirms the prediction of Peyre [96] in this case.

type	# lines	singularity
i	7	$\mathbf{A}_1$
ii	5	$2\mathbf{A}_1$
iii	4	$\mathbf{A}_2$
iv	3	$\mathbf{A}_1 + \mathbf{A}_2$
v	2	$\mathbf{A}_3$
vi	1	$\mathbf{A}_4$

Table 2.7: Classification (over $\overline{\mathbb{Q}}$) of singular del Pezzo surfaces of degree 5

Following [47, Proposition 8.4], Table 2.7 gives the six possible singularity types that arise in the classification of singular del Pezzo surfaces of degree 5. Here, as calculated by Derenthal [53, § 6], the surfaces of type `ii` and `iv` are toric. The surfaces of type `v` and `vi` are equivariant compactifications of $\mathbb{G}_a^2$. It would be interesting to determine whether either of the remaining surfaces arise as equivariant compactifications of further algebraic groups.

2.6 Universal torsors

All proofs of the Manin conjecture seem to have proceeded via one of two routes. When there is a large group action on the variety $V \subseteq \mathbb{P}^{n-1}$ then an analysis of the *height zeta function*

$$Z_U(s) := \sum_{x \in U(\mathbb{Q})} \frac{1}{H(x)^s},$$

is particularly fruitful. The zeta function is defined for $s \in \mathbb{C}$ with $\Re e(s)$ sufficiently large to ensure the absolute convergence of the sum. Theorem 1.2 ensures that it converges for $\Re e(s) > n$. One can pass between analytic information about $Z_U(s)$ and asymptotic behaviour of the corresponding counting function $N_U(B)$ via Tauberian theorems, such as that found below in Lemma 5.5, for example. In fact it follows from Lemma 5.4 that

$$N_U(B) = \frac{1}{2\pi i}\int_{c-i\infty}^{c+i\infty} Z_U(s)\frac{B^s}{s}\mathrm{d}s,$$

where c is arbitrary but exceeds the abscissa of absolute convergence for $Z_U(s)$.

The alternative method to have emerged in the recent past relies upon lifting the counting problem to the *universal torsor*, a certain auxiliary variety parametrising the rational points on U. Universal torsors were originally introduced by Colliot-Thélène and Sansuc [43, 44] to aid in the study of the Hasse principle and weak approximation for rational varieties. Since their inception it is now well-recognised that they also have a central rôle to play in proofs of the Manin conjecture for Fano varieties, and in particular, for del Pezzo surfaces. This point of view was introduced by Salberger [102] in the case of toric varieties. In the case of non-singular hypersurfaces $X \subset \mathbb{P}^{n-1}$, with $n \geqslant 5$, this is exactly the passage from rational vectors $\mathbf{x} = (x_1, \dots, x_n)$, modulo the diagonal action of $\mathbb{Q}$, to primitive lattice points $(\mathbb{Z}^n \setminus \{\mathbf{0}\})/\pm$. An inspection of the proof reveals that this approach was already at the heart of our proof of Theorem 1.2, for example.

Let $S \subset \mathbb{P}^d$ be a del Pezzo surface of degree $d \in \{3,4,5,6\}$, and let $\widetilde{S}$ denote the minimal desingularisation of S if it is singular, with $\widetilde{S} = S$ otherwise. Let $E_1, \dots, E_{10-d} \in \mathrm{Div}(\widetilde{S})$ be generators for $\mathrm{Pic}_{\overline{\mathbb{Q}}}(\widetilde{S})$, and let $E_i^\times = E_i \setminus \{\text{zero section}\}$. Working over $\overline{\mathbb{Q}}$, a *universal torsor* above $\widetilde{S}$ is given by the action of the torus $\mathbb{G}_m^{10-d}$ on the map

$$\pi : E_1^\times \times_{\widetilde{S}} \cdots \times_{\widetilde{S}} E_{10-d}^\times \to \widetilde{S}.$$

A proper discussion of universal torsors would take us too far afield. Instead the reader may consult the survey of Peyre [97] for further details.

Given the usual open subset $U \subset S$, the general theory of universal torsors ensures that there is a partition of $U(\mathbb{Q})$ into a disjoint union of patches, each of which is in bijection with a suitable set of integral points on a universal torsor above $\widetilde{S}$. The explicit determination of equations for universal torsors is a topic of increasing interest in algebraic geometry, involving tools from invariant theory and toric geometry. In this vein there is the construction of Hassett and Tschinkel [67], which outlines an alternative approach to universal torsors via the Cox ring. This method has been developed further by Derenthal [53], in his classification of split del Pezzo surfaces whose universal torsor arises as a hypersurface.

The guiding principle behind the use of universal torsors is simply that they ought to be arithmetically simpler than the original variety. For example, there are many situations in which universal torsors are known to satisfy the Hasse principle and weak approximation, even if the underlying variety fails in this respect. In our work it will suffice to think of universal torsors as "particularly nice parametrisations" of rational points on the surface. The universal torsors that we encounter all have embeddings as affine hypersurfaces of high dimension. Moreover, in each case, we will show how the underlying equation of the universal torsor can be deduced in a completely elementary fashion, without any recourse to geometry. The torsor equations we will meet all take the shape

$$A + B + C = 0,$$

for monomials A, B, C of various degrees in the appropriate variables. As in many examples of counting problems for higher-dimensional varieties, one can occasionally gain leverage by fixing some of the variables at the outset, in order to be left with a counting problem for a family of small-dimensional varieties. If one is sufficiently clever about which variables to fix first, one is sometimes left with a quantity that we know how to estimate — and crucially — whose error term we can control once summed over the remaining variables.

In general terms there are three basic steps in proving the Manin conjecture for split del Pezzo surfaces S:

(i) construct an explicit bijection between rational points of bounded height on S and integral points in a region on the universal torsor above S;

(ii) use analytic number theory to approximate the number of integral points in this region by its volume; and

(iii) show that the volume of this region grows asymptotically in the way that is predicted by the Manin conjecture.

Derenthal [54] has attempted to harmonise many of the basic analytic tools that typically appear in the second step, in order to facilitate future applications of the torsor approach to del Pezzo surfaces.

As a concrete example, we note that Hassett and Tschinkel [67] have calculated the universal torsor for the cubic surface (2.13). It is shown that there is a unique universal torsor above the minimal desingularisation $\widetilde{S}_1$. It is given by the affine equation

$$y_\ell s_\ell^3 s_4^2 s_5 + y_2^2 s_2 + y_1^3 s_1^2 s_3 = 0,$$

for variables $y_1, y_2, y_\ell, s_1, s_2, s_3, s_\ell, s_4, s_5, s_6$. One of the variables does not explicitly appear in this equation, and the torsor should be thought of as being embedded in $\mathbb{A}^{10}$. It turns out that the way to proceed here is to fix all of the variables apart from y_1, y_2, y_ℓ. One may then view the equation as a congruence

$$y_2^2 s_2 \equiv -y_1^3 s_1^2 s_3 \pmod{s_\ell^3 s_4^2 s_5},$$

in order to take care of the summation over y_ℓ. This is the approach taken in [19], the next step being to employ very standard facts about the number of integer solutions to polynomial congruences that are restricted to lie in certain regions.

One is then left with a main term and an error term, which the remaining variables need to be summed over. While the treatment of the main term is relatively routine, the treatment of the error term presents a much more serious obstacle, requiring a wide range of sophisticated estimates for exponential sums.

The universal torsors that turn up in the proofs of Theorems 2.1 and 2.4 can also be embedded in affine space as hypersurfaces. We will see in Chapter 5 that the approach discussed above also produces results for the del Pezzo surface of degree 6 considered in Theorem 2.4. In the proof of Theorem 2.1 in Chapter 6 our approach will be more obviously geometric, and we will actually view the equation as a family of projective lines and also as a family of conics. We will then call upon techniques from the geometry of numbers to count the relevant solutions.

Exercises for Chapter 2

Exercise 2.1. Establish (2.4) when $d = 2$ and V is given by $x_1^2 + x_2^2 = x_3^2$.

Exercise 2.2. The *Hessian* of a hypersurface $f = 0$ in $\mathbb{P}^{n-1}$ is defined to be the determinant of the matrix

$$\left(\frac{\partial^2 f}{\partial x_i \partial x_j}\right)_{1 \leqslant i,j \leqslant n}.$$

Calculate the Hessian of the cubic surfaces (2.14) and (2.15). Conclude that they are not isomorphic over $\overline{\mathbb{Q}}$.

Exercise 2.3. Let $S \subset \mathbb{P}^4$ denote the surface

$$\begin{cases} x_1x_2 - x_3x_4 = 0, \\ x_1^2 + x_2^2 + x_3^2 - x_4^2 - 2x_5^2 = 0. \end{cases}$$

Use the Jacobian criterion to show that S is non-singular. Show explicitly that S has Segre symbol $(1, 1, 1, 1, 1)$.

Exercise 2.4. Calculate the Segre symbol for each of the surfaces in Table 2.4, and check they match up with the correct singularity type in Table 2.3.

Exercise 2.5. Show that $N_U(B) = O_\varepsilon(B^{1+\varepsilon})$ for the surfaces of type `ix`, `x` and `xiv` in Table 2.4.

Chapter 3

The dimension growth conjecture

For any $n \geqslant 3$, let $V \subset \mathbb{P}^{n-1}$ be an irreducible variety of degree d whose ideal is generated by forms defined over the rationals. In this degree of generality one might still ask whether anything meaningful can be said about the corresponding counting function $N_V(B)$, as defined in (1.6). In contrast to the preceding chapter, where precise asymptotic formulae were sought for $N_U(B)$ for suitable open subsets $U \subseteq V$, the aim of the present chapter is to seek general upper bounds for the full counting function $N_V(B)$, making as few assumptions on V as possible.

When $d = 1$, so that V is a linear space, one has

$$N_V(B) \sim c_V B^{\dim V + 1}$$

as $B \to \infty$, for an appropriate constant $c_V > 0$. It is actually rather straightforward to show that $N_V(B)$ can never grow any faster than it does for linear varieties, as the following well-known result demonstrates.

Theorem 3.1. *Let $V \subset \mathbb{P}^{n-1}$ be a variety of degree d. Then we have*

$$N_V(B) \ll_{d,n} B^{\dim V + 1}.$$

Proof. Let us write m for the dimension of V. Let $\hat{V} \subset \mathbb{A}^n$ be the affine cone above V. Then $\hat{V}$ is an affine variety of degree d and dimension $m + 1$. For any affine variety $T \subset \mathbb{A}^\nu$ of degree δ and dimension μ, we define the quantity

$$M_T(B) := \#\{\mathbf{t} \in T \cap \mathbb{Z}^\nu : |\mathbf{t}| \leqslant B\}.$$

We will show that

$$M_T(B) = O_{\delta,\nu}(B^\mu). \tag{3.1}$$

This will clearly suffice to establish the theorem, since $N_V(B) \leqslant M_{\hat{V}}(B)$.

We establish (3.1) by induction on μ. Since a variety of dimension zero and degree δ contains δ points, the estimate is trivial when $\mu = 0$. Assume that $\mu \geqslant 1$. It will suffice to assume that T is irreducible, since T clearly contains at most δ such components. It therefore follows that we can find an index $1 \leqslant a \leqslant \nu$ such that T intersects the hyperplane $t_a = \alpha$ properly, for any $\alpha \in \mathbb{C}$. Let H_α denote this hyperplane. Then we have

$$M_T(B) \leqslant \sum_{|\alpha| \leqslant B} M_{T \cap H_\alpha}(B).$$

Since $T \cap H_\alpha$ has dimension at most $\mu - 1$ for every α, and decomposes into at most δ irreducible components, an application of the induction hypothesis implies that $M_{T \cap H_\alpha}(B) = O_{\delta,\nu}(B^{\mu-1})$. This suffices to complete the proof of (3.1), and so completes the proof of Theorem 3.1. □

We will henceforth think of Theorem 3.1 as providing the "trivial bound" for the counting function, and we would like to know to what extent it can be improved upon. Clearly some additional assumptions are necessary here, as we have already seen that this bound is optimal when V contains a linear component of maximal dimension that is defined over the rationals. Likewise, whenever V contains a linear divisor which is defined over the rationals it follows that

$$N_V(B) \gg_V B^{\dim V}.$$

We met with a hypersurface of this sort when we considered the form (1.5).

Recall the definition of β_V from (2.1). The following conjecture, made by Heath-Brown [74, Conjecture 1] in the case of hypersurfaces, predicts that this lower bound should be matched by an upper bound of similar quality.

Conjecture 3.1. *Let $V \subset \mathbb{P}^{n-1}$ be an irreducible variety of degree $d > 1$. Then we have $\beta_V \leqslant \dim V$.*

Once formulated in terms of the associated counting function, Conjecture 3.1 therefore has the following consequence.

Conjecture 3.2 (Dimension growth). *Let $V \subset \mathbb{P}^{n-1}$ be an irreducible variety of degree $d > 1$. Then we have*

$$N_V(B) \ll_{\varepsilon,V} B^{\dim V + \varepsilon},$$

for any $\varepsilon > 0$.

Heath-Brown [74, Conjecture 2] goes further, predicting that the upper bound in Conjecture 3.2 should hold with an implied constant that is allowed to depend at most upon d and n, in addition to the choice of ε.

Conjecture 3.3 (Uniform dimension growth). *Let $V \subset \mathbb{P}^{n-1}$ be an irreducible variety of degree $d > 1$. Then we have*

$$N_V(B) \ll_{d,n,\varepsilon} B^{\dim V + \varepsilon},$$

for any $\varepsilon > 0$.

It turns out that for many applications it is very useful to have uniformity in the coefficients of the polynomials defining the variety V. This makes Conjecture 3.3 worth the effort of pursuing. In our discussion of the dimension growth conjectures, however, we will place emphasis on the weaker Conjecture 3.2, in which an arbitrary dependence on the coefficients is allowed.

3.1 Linear spaces on hypersurfaces

Several approaches to the dimension growth conjecture have involved paying attention to the linear spaces that are allowed to lie in the relevant varieties. For non-negative integers $m \leqslant N$, let $\mathbb{G}(m,N)$ denote the *Grassmannian variety* which parametrises m-planes contained in $\mathbb{P}^N$. Here, an *m-plane* is a linear space of dimension m. When $m = N-1$ we sometimes write $\mathbb{P}^{N^*} \cong \mathbb{P}^N$ for the space of hyperplanes in $\mathbb{P}^N$. For general m it is well-known that $\mathbb{G}(m,N)$ can be embedded in $\mathbb{P}^\nu$ via the *Plücker embedding*, where

$$\nu = \binom{N+1}{m+1} - 1,$$

with

$$\dim \mathbb{G}(m,N) = (m+1)(N-m). \tag{3.2}$$

If $M \in \mathbb{G}(m,N)(\mathbb{Q}) = \mathbb{G}(m,N) \cap \mathbb{P}^\nu(\mathbb{Q})$, we define the height $H(M)$ of M to be the standard exponential height of its coordinates in $\mathbb{G}(m,N)$, under the Plücker embedding. The following result, which amounts to a simple application of the geometry of numbers, will be established in Section 4.2 as a consequence of Lemma 4.5.

Lemma 3.1. *Let $M \in \mathbb{G}(m,N)(\mathbb{Q})$. Then we have*

$$N_M(B) \ll_N \frac{B^{m+1}}{H(M)} + \begin{cases} 1, & \text{if } m = 1, \\ B^m, & \text{if } m \geqslant 2. \end{cases}$$

Now let $V \subset \mathbb{P}^{n-1}$ be an irreducible variety of degree d and let $m \in \mathbb{N}$. The variety parametrising all m-planes contained in V is

$$F_m(V) := \{\Lambda \in \mathbb{G}(m, n-1) : \Lambda \subseteq V\}. \tag{3.3}$$

This is usually called the *Fano variety* parametrising m-planes. For example, if $V \subset \mathbb{P}^3$ is a non-singular cubic surface, then we saw in the previous chapter that $F_1(V)$ is zero-dimensional of degree 27.

The following estimate may certainly be extracted from the work of Segre [110], although the proof is so straightforward that we have included it here.

Lemma 3.2. *Let $V \subset \mathbb{P}^{n-1}$ be an irreducible variety of degree d. Then we have*

$$\dim F_1(V) \leqslant \begin{cases} 2\dim V - 2, & \text{if } d = 1, \\ 2\dim V - 3, & \text{if } d \geqslant 2. \end{cases}$$

Proof. Let $\delta = \dim V$. In order to establish Lemma 3.2 we note that the case in which V is isomorphic to $\mathbb{P}^\delta$ is easy, since then $F_1(V) \cong \mathbb{G}(1,\delta)$ and the result follows from (3.2). Assuming therefore that $d \geqslant 2$, we employ a routine incidence correspondence argument. Let Z be an irreducible component of $F_1(V)$, and let

$$\Sigma = \{(v, L) \in V \times Z : \ v \in L\}.$$

By considering the projection onto the second factor we deduce that $\dim \Sigma = \dim Z + 1$. Now let $V_0 \subseteq V$ be the union of the lines in Z, and let v be a generic point of V_0. The projection $\Sigma \to V$ then shows that $\dim \Sigma = \dim V_0 + \dim Z_v$, where $Z_v = \{L \in Z : v \in L\}$. Thus

$$\dim Z = \dim \Sigma - 1 = \dim V_0 - 1 + \dim Z_v \leqslant \dim V - 1 + \dim Z_v.$$

Now any line $L \in Z_v$ must also lie in the tangent space $\mathbb{T}_v(V_0)$, so that $Z_v \subseteq W_v$, where

$$W_v = \{L \in \mathbb{G}(1, N) : v \in L \subseteq \mathbb{T}_v(V_0)\}.$$

Since v is generic on V_0, it is non-singular, so that W_v is a linear space of dimension $\dim V_0 - 1$.

We proceed to consider two cases. If V_0 is a linear space, then it must be a proper subvariety of V, since $d \geqslant 2$. In this case

$$\dim Z_v \leqslant \dim W_v = \dim V_0 - 1 \leqslant \dim V - 2,$$

and the required result follows. On the other hand, if V_0 is not linear, then Z_v must be a proper subvariety of W_v, and

$$\dim Z_v \leqslant \dim W_v - 1 = \dim V_0 - 2 \leqslant \dim V - 2.$$

Again this suffices for the lemma. □

In the case of cubic surfaces in $\mathbb{P}^3$ Lemma 3.2 says that the Fano variety of lines is at most 1-dimensional. This is off the mark for non-singular surfaces, as discussed above, but is attained when the cubic surface is actually a cone over an irreducible plane cubic curve. A similar argument provides the following analogue for planes contained in a variety.

Lemma 3.3. *Let $V \subset \mathbb{P}^{n-1}$ be an irreducible variety of degree d. Then we have*

$$\dim F_2(V) \leqslant \begin{cases} 3\dim V - 6, & \text{if } d = 1, \\ 3 \dim V - 8, & \text{if } d \geqslant 2. \end{cases}$$

For the remainder of this section we let $X \subset \mathbb{P}^{n-1}$ be a hypersurface of degree $d > 1$. We will always assume that it is irreducible. Now it is obvious that $F_m(X) = \emptyset$ for $m \geqslant n-2$. When X is assumed to be non-singular, one can extend the range for m in this inequality.

Lemma 3.4. *Assume that $X \subset \mathbb{P}^{n-1}$ is a non-singular hypersurface of degree $d > 1$. Then $F_m(X) = \emptyset$ if $m > \frac{n-1}{2}$.*

Proof. Let $m > \frac{n-1}{2}$ be an integer and let $\Lambda \in F_m(X)$. After a change of variables we may assume without loss of generality that Λ is given by $x_{m+1} = \cdots = x_n = 0$. Let F be the underlying form that defines X. Since $\Lambda \subset X$ we have $F(x_1, \dots, x_m, 0, \dots, 0)$ being identically zero. Likewise,

$$\frac{\partial F}{\partial x_i}(x_1, \dots, x_m, 0, \dots, 0)$$

is identically zero for $1 \leqslant i \leqslant m$. In view of the fact that $d > 1$, it follows that the homogeneous polynomial

$$\frac{\partial F}{\partial x_{m+i}}(x_1, \dots, x_m, 0, \dots, 0)$$

is non-constant for $1 \leqslant i \leqslant n-m$. Since $n-1-m < m$, these $n-1-m$ non-constant homogeneous polynomials have a common zero in Λ. By the Jacobian criterion, this is therefore a singular point of X. Hence it follows that when $2m > n-1$, every m-plane $\Lambda \subset X$ intersects the singular locus of X. In particular $F_m(X)$ is empty when X is non-singular. □

Lemma 3.4 leads one to ask whether non-singular hypersurfaces can contain m-planes with $m = \frac{n-1}{2}$ and n odd. This does in fact arise, as the example of the Fermat hypersurface

$$x_1^d + \cdots + x_n^d = 0$$

demonstrates. This variety contains the obvious linear space

$$x_1 + x_2 = x_3 + x_4 = \cdots = x_{n-1} + x_n = 0$$

when d is odd. Nonetheless, it has been shown by Starr [33, Appendix] that when $X \subset \mathbb{P}^{n-1}$ is non-singular and n is odd, then it contains at most finitely many $(\frac{n-1}{2})$-planes.

When m is small, there are some further results on the possible dimension of $F_m(X)$ when X is non-singular. In the case $m = 1$ of lines, a standard incidence correspondence argument (see Harris [65, § 12.5], for example) reveals that $\dim F_1(X) = 2n-5-d$ for a generic hypersurface $X \subset \mathbb{P}^{n-1}$ of degree d. Debarre and de Jong have conjectured this to be the true dimension of $F_1(X)$ whenever $d \leqslant n-1$. The following result, due to Beheshti [4], makes some progress in this direction.

Lemma 3.5. *Assume that $X \subset \mathbb{P}^{n-1}$ is a non-singular hypersurface of degree $d > 1$. Then $\dim F_1(X) = 2n-5-d$ if $d \leqslant \min\{6, n-1\}$.*

For comparison, we note that an application of Lemma 3.2 would only yield $\dim F_1(X) \leqslant 2n-7$. Note that in Lemma 3.5 we retrieve the fact that $F_1(X)$ has dimension 0 when $d = 3$ and $n = 4$.

When $m = 2$, the situation is less well understood. While the dimension of $F_2(X)$ is $3n - 9 - \frac{1}{2}(d+1)(d+2)$ for a generic hypersurface $X \subset \mathbb{P}^{n-1}$ of degree d, there exist examples showing that this is not always the true dimension when $d \leqslant n-1$. Perhaps the simplest example is provided by the Fermat cubic in $\mathbb{P}^5$, which contains a finite number of planes.

3.2 Dimension growth for hypersurfaces

By using an argument involving birational projections, as in [36, Corollary 1], it is possible to show that Conjectures 3.2 and 3.3 are equivalent to the corresponding conjectures in which V is replaced by an arbitrary irreducible hypersurface of degree exceeding 1. Let us suppose, therefore, that $X \subset \mathbb{P}^{n-1}$ is an irreducible hypersurface of degree $d > 1$. Assuming that X is irreducible, Conjecture 3.2 has the following consequence, since $\dim X = n-2$.

Conjecture 3.4. *Let $X \subset \mathbb{P}^{n-1}$ be an irreducible hypersurface of degree $d > 1$. Then we have*

$$N_X(B) \ll_{\varepsilon,X} B^{n-2+\varepsilon},$$

for any $\varepsilon > 0$.

There are two essential aspects in which Conjecture 3.4 is essentially best possible. Firstly, when X is defined by a non-singular quadratic form $Q \in \mathbb{Z}[x_1, \ldots, x_n]$, with $n \geqslant 3$, one can show that

$$N_X(B) \sim c_Q B^{n-2} (\log B)^{\delta_Q}$$

as $B \to \infty$. Here, $c_Q \geqslant 0$ is a product of local densities which is positive if and only if Q is indefinite and it represents zero non-trivially modulo p^k, for every prime power p^k. Furthermore, $\delta_Q \in \{0, 1\}$, with $\delta_Q = 1$ if and only if $n = 4$ and the discriminant of Q is a square. In particular $N_X(B) \gg_Q B^{n-2}$ for all quadrics defined by isotropic quadratic forms in $n \geqslant 3$ variables. These facts can be proved using the Hardy–Littlewood circle method, a technique in analytic number theory that will be examined in Chapter 8.

Suppose now that X is defined by a polynomial of the shape (1.5), or more generally,

$$x_1 F_1(\mathbf{x}) - x_2 F_2(\mathbf{x}),$$

with $F_1, F_2 \in \mathbb{Z}[x_1, \ldots, x_n]$ forms of degree $d-1$ such that $x_1 F_1 - x_2 F_2$ is absolutely irreducible. Then in view of the trivial solutions with $x_1 = x_2 = 0$, which simply correspond to the obvious linear divisor contained in X, one has the lower bound $N_X(B) \gg B^{n-2}$. One might ask when linear subspaces can arise in this way. It follows from Lemma 3.4 that situations like the above cannot arise for non-singular hypersurfaces $X \subset \mathbb{P}^{n-1}$ of degree $d > 1$ when $n \geqslant 6$. In fact, when $d \geqslant 3$, Conjecture 2.1 predicts that $N_X(B) = O_X(B^{n-2-\delta})$ for some $\delta > 0$.

Returning to the setting of arbitrary hypersurfaces of degree $d > 1$, we have already seen that Conjecture 3.4 holds when $d = 2$. In Table 3.1 we have charted the progress that has been made towards Conjecture 3.4 in the remaining cases. In doing so we have maintained a rough chronological order of discovery. The table presents values of $\theta_{d,n} \in \mathbb{R}$ for which an upper bound of the form

$$N_X(B) \ll_{\varepsilon,X} B^{n-2+\theta_{d,n}+\varepsilon},$$

holds for irreducible hypersurfaces $X \subset \mathbb{P}^{n-1}$ of degree $d \geqslant 3$. We will write σ_X for the projective dimension of the singular locus of X. Thus σ_X takes values in the set $\{-1, 0, \ldots, n-2\}$, with $\sigma_X = -1$ if and only if X is non-singular.

	θ_d	Restrictions?	Who?
(i)	$\frac{1}{2}$	—	Cohen [112]
(ii)	$\frac{2}{n}$	$\sigma_X = -1$	Fujiwara [61]
(iii)	$\frac{15}{n+5} - 1$	$\sigma_X = -1$	Heath-Brown [70]
(iv)	$\frac{1}{d}$	—	Pila [100]
(v)	0	$n \leqslant 4$	Heath-Brown [74]
(vi)	0	$n = 5$	Browning & Heath-Brown [30]
(vii)	0	$d \geqslant 6$	Browning, Heath-Brown & Salberger [36]
(viii)	0	$\sigma_X = -1$	Browning & Heath-Brown [32, 33]
(ix)	0	$d \geqslant 4$	Salberger [106]
(x)	0	$d = 3$ and $n \geqslant 6 + \sigma_X$	Browning [24]

Table 3.1: Progress on the dimension growth conjecture

The methods used to prove these estimates are quite varied, encompassing sieve methods, exponential sums, the geometry of numbers and ideas stemming from the determinant method of Bombieri–Pila [7] and its extension by Heath-Brown [74]. Note that Cohen's result remains unpublished, but the reference given is to an alternative proof by Serre using the large sieve inequality. In Table 3.1 items (i)–(iii) are proved using exponential sums. We will discuss this sort of approach in Section 3.3 below. Items (iv), (v), (vii) and (ix) are based on the determinant method that is discussed in Section 4.1. Items (vi) and (viii) use the geometry of numbers in an essential way. Some of the ideas involved are presented in Section 3.4. Finally, item (x) is proved using the Hardy–Littlewood circle method, a taste of which can be gleaned from Chapter 8.

There has clearly been a lot of work on the dimension growth conjecture in recent times. We are now in the happy position of being left to deal with singular irreducible hypersurfaces $X \subset \mathbb{P}^{n-1}$ of degree $d = 3$ and dimension $n-2$, with

$$4 \leqslant n - 2 < 4 + \sigma_X, \quad \sigma_X \geqslant 0.$$

It remains an interesting and difficult challenge to complete the proof of the dimension growth conjecture for cubic hypersurfaces.

3.3 Exponential sums

In this section we show how the theory of exponential sums can be useful in the context of the dimension growth conjecture. Let $X \subset \mathbb{P}^{n-1}$ be a hypersurface, defined by a form $F \in \mathbb{Z}[x_1, \ldots, x_n]$. Rather surprisingly, a very profitable approach to bounding $N_X(B)$ is based upon the trivial observation that

$$N_X(B) \leqslant \frac{1}{2} N(F; B, m)$$

for any $m \in \mathbb{N}$, where

$$N(F; B, m) := \#\{\mathbf{x} \in \mathbb{Z}^n : m \mid F(\mathbf{x}), \ |\mathbf{x}| \leqslant B\}.$$

In practice one chooses $m = p$, for some large prime p. The approach in (iii) of Table 3.1 is based instead on van der Corput's method for exponential sums, initiated by working with $m = pq$ for distinct large primes p and q. A refinement of the method can be found in recent work of Marmon [92]. In this section we will discuss a simple-minded application of exponential sums which leads to the following result.

Theorem 3.2. *Assume that $X \subset \mathbb{P}^{n-1}$ is irreducible of degree at least 2, with $n \geqslant 3$. Then we have*

$$N_X(B) \ll_X \begin{cases} B^{n-2+\frac{2}{n+1}} (\log B)^{\frac{2}{n-1}}, & \text{if } X \text{ non-singular,} \\ B^{n-1-\frac{1}{n-1}} (\log B)^{\frac{2}{n-1}}, & \text{otherwise.} \end{cases}$$

In fact the implied constant in both of these bounds can be made to depend at most on n and the degree of X. The second estimate is weaker than any of the bounds presented in Table 3.1 but does not involve any restrictions on the singular locus of the hypersurface involved. Exercise 3.4 records a sharpening of the exponent of B to $n - 1 - \frac{3}{2n-3} + \varepsilon$. The first bound in Theorem 3.2 is due to Heath-Brown [70, Theorem 1] and improves slightly on entry (ii) in the table.

Once the translation to exponential sums over finite fields has been achieved, the key underlying technology will be the estimates of Lang–Weil [88] and Deligne [51] for the number of points on affine varieties defined over finite fields. The latter requires deep machinery from algebraic geometry and is outside the scope of this book. The former is straightforward and yields the following result, whose proof is Exercise 3.3.

Lemma 3.6. *Let $V \subset \mathbb{A}^n$ be an irreducible affine variety of degree $d \geqslant 2$, defined over a finite field $\mathbb{F}_p$. Then we have*

$$\#V(\mathbb{F}_p) \ll_{d,n} p^{\dim V}.$$

In fact, Lang and Weil go somewhat further than the crude upper bound in Lemma 3.6, showing in particular that

$$\#V(\mathbb{F}_p) = p^{\dim V} + O_{d,n}(p^{\dim V - \frac{1}{2}}), \tag{3.4}$$

under the same hypotheses. The Deligne estimates are stronger still, but require the additional hypothesis that the affine variety V is non-singular modulo p. We will only require the following version, in which V is taken to be an affine hypersurface.

Lemma 3.7. *Let $V \subset \mathbb{A}^n$ be an affine hypersurface of degree $d \geqslant 2$, defined over a finite field $\mathbb{F}_p$. Suppose that the homogeneous degree d part of the underlying polynomial is non-singular. Then we have*

$$\#V(\mathbb{F}_p) = p^{n-1} + O_{d,n}(p^{\frac{n}{2}}).$$

We now have all the tools in place with which to carry out the proof of Theorem 3.2, during which we will closely follow the argument of Heath-Brown [70]. It will be convenient to allow all of the ensuing implied constants to depend implicitly on the hypersurface X. We will adopt the usual notation $e(\cdot) = e^{2\pi i \cdot}$ and $e_p(\cdot) = e(\frac{\cdot}{p})$ for any prime p.

Consider the function

$$w(\mathbf{z}) := \prod_{i=1}^{n} \left(\frac{\sin \pi z_i}{\pi z_i}\right)^2.$$

This has Fourier transform

$$\hat{w}(\mathbf{y}) := \int_{\mathbb{R}^n} w(\mathbf{z}) e(-\mathbf{y}.\mathbf{z}) \mathrm{d}\mathbf{z} = \prod_{i=1}^{n} \max\{1 - |y_i|, 0\}. \tag{3.5}$$

Let p be a prime. Then, if F denotes the underlying form that defines the hypersurface X, our first step in the proof of Theorem 3.2 is the observation that

$$N_X(B) \leqslant \frac{1}{2} N(F; B, p) \ll \sum_{\substack{\mathbf{x} \in \mathbb{Z}^n \\ p \mid F(\mathbf{x})}} w\Big(\frac{\mathbf{x}}{2B}\Big).$$

Here, as in the remainder of the proof, we will allow our implied constants to depend on X. An application of the Poisson summation formula reveals that

$$\begin{aligned} \sum_{\substack{\mathbf{x} \in \mathbb{Z}^n \\ p \mid F(\mathbf{x})}} w\Big(\frac{\mathbf{x}}{2B}\Big) &= \sum_{\substack{\mathbf{y} \in \mathbb{F}_p^n \\ p \mid F(\mathbf{y})}} \sum_{\mathbf{z} \in \mathbb{Z}^n} w\Big(\frac{\mathbf{y} + p\mathbf{z}}{2B}\Big) \\ &= \sum_{\substack{\mathbf{y} \in \mathbb{F}_p^n \\ p \mid F(\mathbf{y})}} \sum_{\mathbf{c} \in \mathbb{Z}^n} \int_{\mathbb{R}^n} w\Big(\frac{\mathbf{y} + p\mathbf{t}}{2B}\Big) e(-\mathbf{c}.\mathbf{t}) \mathrm{d}\mathbf{t}. \end{aligned}$$

Making the substitutions $\mathbf{s} = \mathbf{y} + p\mathbf{t}$ and then $\mathbf{r} = \frac{\mathbf{s}}{2B}$, we therefore find that

$$\begin{aligned}\sum_{\substack{\mathbf{x}\in\mathbb{Z}^n\\ p\mid F(\mathbf{x})}} w\Big(\frac{\mathbf{x}}{2B}\Big) &= \frac{1}{p^n}\sum_{\substack{\mathbf{y}\in\mathbb{F}_p^n\\ p\mid F(\mathbf{y})}}\sum_{\mathbf{c}\in\mathbb{Z}^n} e_p(\mathbf{c}.\mathbf{y})\int_{\mathbb{R}^n} w\Big(\frac{\mathbf{s}}{2B}\Big)e_p(-\mathbf{c}.\mathbf{s})\mathrm{d}\mathbf{s}\\ &= \frac{2^nB^n}{p^n}\sum_{\substack{\mathbf{y}\in\mathbb{F}_p^n\\ p\mid F(\mathbf{y})}}\sum_{\mathbf{c}\in\mathbb{Z}^n} e_p(\mathbf{c}.\mathbf{y})\hat{w}\Big(\frac{2B\mathbf{c}}{p}\Big).\end{aligned}$$

In view of (3.5), this yields the upper bound

$$N_X(B) \ll \frac{B^n}{p^n}\sum_{\substack{\mathbf{c}\in\mathbb{Z}^n\\ |\mathbf{c}|\ll \frac{p}{B}}} |S_p(\mathbf{c})|,$$

where

$$S_p(\mathbf{c}) := \sum_{\substack{\mathbf{y}\in\mathbb{F}_p^n\\ p\mid F(\mathbf{y})}} e_p(\mathbf{c}.\mathbf{y}).$$

We shall henceforth make the assumption that the prime p lies in the interval $P \leqslant p < 2P$, with $P \gg B$. Write $\mathscr{S}_1$ for the set of $\mathbf{c} \in \mathbb{Z}^n$ for which $|\mathbf{c}| \ll \frac{P}{B}$ and $p \nmid \mathbf{c}$ and write $\mathscr{S}_2$ for the set of $\mathbf{c} \in \mathbb{Z}^n$ such that $|\mathbf{c}| \ll \frac{P}{B}$ and $p \mid \mathbf{c}$. Since $\#\mathscr{S}_2 = O(1)$ it easily follows that

$$N_X(B) \ll \frac{B^n}{P} + \frac{B^n}{P^n}\sum_{\mathbf{c}\in\mathscr{S}_1} |S_p(\mathbf{c})|, \tag{3.6}$$

by Lemma 3.6. In order to handle the second term in (3.6) we replace $\mathbf{y}$ by $a\mathbf{y}$ for each non-zero $a \pmod p$, finding that

$$\begin{aligned}(p-1)S_p(\mathbf{c}) &= \sum_{a=1}^{p-1}\sum_{\substack{a\mathbf{y}\in\mathbb{F}_p^n\\ p\mid F(a\mathbf{y})}} e_p(a\mathbf{c}.\mathbf{y})\\ &= \sum_{\substack{\mathbf{y}\in\mathbb{F}_p^n\\ p\mid F(\mathbf{y})}}\Big(\sum_{a=1}^{p} e_p(a\mathbf{c}.\mathbf{y}) - 1\Big)\\ &= p\#\{\mathbf{y}\in\mathbb{F}_p^n : p \mid F(\mathbf{y}),\ p \mid \mathbf{c}.\mathbf{y}\} - \#\{\mathbf{y}\in\mathbb{F}_p^n : p \mid F(\mathbf{y})\}\\ &= pN_1(p) - N_2(p),\end{aligned}$$

say. Thus it follows that

$$S_p(\mathbf{c}) \ll \frac{1}{p}|pN_1(p) - N_2(p)|. \tag{3.7}$$

Our work now diverges according to whether or not X is non-singular.

3.3.1 Singular X

Beginning with the case in which no assumptions are made on the singular locus of X, we merely assume that F is absolutely irreducible. Let $\mathbf{c} \in \mathscr{S}_1$. Since $p \nmid \mathbf{c}$, we may assume without loss of generality that $p \nmid c_n$. Hence it is possible to eliminate the variable y_n from the pair of congruences in $N_1(p)$. This leads to the conclusion that

$$N_1(p) = \#\{\mathbf{y}' \in \mathbb{F}_p^{n-1} : p \mid F_{\mathbf{c}}(\mathbf{y}')\},$$

where $\mathbf{y}' = (y_1, \ldots, y_{n-1})$ and $F_{\mathbf{c}}$ is the form obtained by eliminating y_n from F. Now either $F_{\mathbf{c}}$ vanishes identically, or we have $N_1(p) \ll p^{n-2}$ by Lemma 3.6.

Write $\mathscr{S}_1 = \mathscr{S}_{1,a} \cup \mathscr{S}_{1,b}$, where $\mathscr{S}_{1,a}$ denotes the set of $\mathbf{c} \in \mathscr{S}_1$ for which $F_{\mathbf{c}}$ does not vanish identically modulo p, and $\mathscr{S}_{1,b} = \mathscr{S}_1 \setminus \mathscr{S}_{1,a}$. We conclude from (3.6) and (3.7) that

$$N_X(B) \ll \frac{B^n}{P} + P^{n-2} + \frac{B^n}{P^n} \sum_{\mathbf{c} \in \mathscr{S}_{1,b}} |S_p(\mathbf{c})|, \tag{3.8}$$

provided that $P \gg B$.

In order to handle the final summation, we write $\mathscr{P} = \mathscr{P}(P; \mathbf{c}, F)$ for the set of primes $P \leqslant p < 2P$ such that the form $F_{\mathbf{c}}$ vanishes identically. In characteristic zero it is a standard geometric fact that the variety obtained from the hyperplane section $F(\mathbf{x}) = \mathbf{c.x} = 0$ will vanish only if $F = 0$ contains the hyperplane $\mathbf{c.x} = 0$. Let $\|F\|$ denote the maximum modulus of the coefficients of F. Since F is assumed to be absolutely irreducible of degree at least 2, so it follows that

$$\#\mathscr{P} \ll \log \|F_{\mathbf{c}}\| \ll \log(P\|F\|) \ll \log P.$$

Here we recall our convention that the implied constants in our proof of Theorem 3.2 are allowed to depend upon the coefficients of F. In view of Lemma 3.6 we clearly have

$$\frac{B^n}{P^n} \sum_{\mathbf{c} \in \mathscr{S}_{1,b}} |S_p(\mathbf{c})| \ll \frac{B^n}{P} \#\{\mathbf{c} \in \mathscr{S}_{1,b} : p \in \mathscr{P}\}.$$

Hence we may average over all primes $P \leqslant p < 2P$ in order to deduce from (3.8) that

$$\begin{aligned} N_X(B) &\ll \frac{B^n}{P} + P^{n-2} + \frac{B^n \log P}{P^2} \sum_{\mathbf{c} \in \mathscr{S}_{1,b}} \#\mathscr{P} \\ &\ll \frac{B^n}{P} + P^{n-2}(\log P)^2. \end{aligned}$$

We therefore complete the proof of the second bound in Theorem 3.2 by choosing $P = B^{\frac{n}{n-1}}(\log B)^{-\frac{2}{n-1}}$.

3.3.2 Non-singular X

In the case of non-singular F we can do much better by applying Lemma 3.7 over Lemma 3.6 in (3.7). To do so we will need to pay attention to how often a hyperplane section of $F = 0$ produces a singular variety. This topic is taken up in Section 7.1. As a consequence of Lemma 7.3 it follows that there exists an absolutely irreducible form $F^* \in \mathbb{Z}[x_1, \ldots, x_n]$ of degree D, with $2 \leqslant D \ll 1$, such that the hyperplane section $F(\mathbf{x}) = \mathbf{c}.\mathbf{x} = 0$ is singular in $\mathbb{F}_p$ if and only if $p \mid F^*(\mathbf{c})$. We proceed by breaking the set $\mathscr{S}_1$ into two sets $\mathscr{S}_{1,a} \cup \mathscr{S}_{1,b}$ in (3.6), where now $\mathscr{S}_{1,a}$ is the set of $\mathbf{c} \in \mathscr{S}_1$ for which $p \nmid F^*(\mathbf{c})$ and $\mathscr{S}_{1,b}$ is the remaining set.

Applying Lemma 3.7 in (3.7) we easily deduce that

$$S_p(\mathbf{c}) \ll \begin{cases} p^{\frac{n-1}{2}}, & \text{if } \mathbf{c} \in \mathscr{S}_{1,a}, \\ p^{\frac{n}{2}}, & \text{if } \mathbf{c} \in \mathscr{S}_{1,b}, \end{cases}$$

provided that F is non-singular modulo p. In particular it now follows from (3.6) that

$$N_X(B) \ll \frac{B^n}{P} + P^{\frac{n-1}{2}} + \frac{B^n}{P^{\frac{n}{2}}} \#\mathscr{S}_{1,b},$$

provided that $P \gg B$ and F is non-singular modulo p. Finally, we average this result over all primes $p \in \mathscr{Q}$, where $\mathscr{Q}$ is the set of primes $P \leqslant p < 2P$ for which F is non-singular modulo p. There are $O(1)$ primes p for which F is singular modulo p. We therefore obtain the bound

$$N_X(B) \ll \frac{B^n}{P} + P^{\frac{n-1}{2}} + \frac{B^n \log P}{P^{\frac{n}{2}+1}} \sum_{p \in \mathscr{Q}} \#\Big\{\mathbf{c} \in \mathbb{Z}^n : |\mathbf{c}| \ll \frac{P}{B},\ p \mid F^*(\mathbf{c})\Big\},$$

provided that $P \gg B$.

Write, temporarily, $P_B = B^{-1}P$. Let $\mathbf{c} \in \mathbb{Z}^n$ be such that $|\mathbf{c}| \ll P_B$. Then a non-zero value of $F^*(\mathbf{c})$ can have at most $O(\log P)$ prime factors, since $F^*(\mathbf{c}) \ll \|F^*\| P^D \ll P^D$. Hence we obtain the estimate

$$\begin{aligned} \sum_{p \in \mathscr{Q}} \#\{\mathbf{c} \in \mathbb{Z}^n : |\mathbf{c}| \ll P_B,\ p \mid F^*(\mathbf{c})\} &\ll \sum_{\substack{\mathbf{c} \in \mathbb{Z}^n \\ |\mathbf{c}| \ll P_B}} \#\{p \in \mathscr{Q} : p \mid F^*(\mathbf{c})\} \\ &\ll \frac{P^n \log P}{B^n} + \frac{P}{\log P} \cdot N(F^*; P_B), \end{aligned}$$

in the notation of (1.2). In view of the fact that F^* is absolutely irreducible and has degree $2 \leqslant D \ll 1$, we may apply the second part of Theorem 3.2, as previously established in Section 3.3.1, to deduce that

$$N(F^*; P_B) \ll P_B^{n-1-\frac{1}{n-1}} (\log P)^{\frac{2}{n-1}}.$$

Combining our various estimates, we may therefore conclude that

$$\begin{aligned} N_X(B) &\ll \frac{B^n}{P} + P^{\frac{n-1}{2}} + \frac{B^n}{P^{\frac{n}{2}+1}}\Big(\frac{P^n(\log P)^2}{B^n} + P\cdot\Big(\frac{P}{B}\Big)^{n-1-\frac{1}{n-1}}(\log P)^{\frac{2}{n-1}}\Big) \\ &\ll \frac{B^n}{P} + P^{\frac{n-1}{2}} + B^{1+\frac{1}{n-1}}P^{\frac{n}{2}-1-\frac{1}{n-1}}(\log P)^{\frac{2}{n-1}}, \end{aligned}$$

provided that $P \gg B$. Taking $P = B^{\frac{2n}{n+1}}$ therefore completes the proof of the first bound in Theorem 3.2.

3.4 Covering with linear spaces

In this section we discuss a completely different approach to the dimension growth conjecture, due to Browning and Heath-Brown [30]. The main tool is a versatile result, valid for hypersurfaces $X \subset \mathbb{P}^{n-1}$ of any dimension, which shows that every point counted by $N_X(B)$ must lie on one of a small number of linear subspaces contained in X, each of relatively low height. As usual, we assume that X is defined by a form $F \in \mathbb{Z}[x_1, \ldots, x_n]$. The following combines a special case of [30, Theorem 4] with [30, Theorem 5].

Theorem 3.3. *Let $X \subset \mathbb{P}^{n-1}$ be an irreducible hypersurface of degree d and let $\varepsilon > 0$. Then there exist linear spaces $M_1, \ldots, M_J \subseteq X$ defined over $\mathbb{Q}$, with $J = O_{d,n,\varepsilon}(B^{n-2+\varepsilon})$, such that the following hold:*

(i) $0 \leqslant \dim M_j \leqslant n-2$ *for* $1 \leqslant j \leqslant J$;

(ii) $H(M_j) = O_n(B(\log \|F\| B)^{n^3})$ *for* $1 \leqslant j \leqslant J$ *such that* $\dim M_j > 0$;

(iii) $\{x \in X(\mathbb{Q}) : H(x) \leqslant B\} \subseteq \bigcup_{j=1}^{J} \{x \in M_j(\mathbb{Q}) : H(x) \leqslant B\}$.

We will see a proof of this result for the case $n = 3$ in Section 4.3. The height of a $\mathbb{Q}$-linear subspace was defined in Section 3.1.

Recall the definition (3.3) of the Fano variety of lines $F_1(X)$ contained in X. Consider the union of lines

$$Y = \bigcup_{L \in F_1(X)} L,$$

which is clearly a subvariety of X. We will say that X is *ruled* if $Y = X$. Recall from Lemma 3.2 that $\dim F_1(X) \leqslant 2n-7$ if X is irreducible and has degree at least 2. Thus, assuming that we are in this setting, with X ruled, it follows that

$$n - 3 \leqslant \dim F_1(X) \leqslant 2n - 7. \tag{3.9}$$

The lower bound follows from the fact that $n - 2 = \dim Y \leqslant 1 + \dim F_1(X)$ when X is ruled.

In the statement of Theorem 3.3 it is clear that the contribution from points lying on linear spaces of dimension 0 (that is to say, from individual points) is satisfactory from the point of view of Conjecture 3.4. Now if $x \in X$ lies on some linear space $M \subseteq X$ of dimension at least 1, then it obviously lies in Y. In particular, Theorem 3.3 implies that the dimension growth conjecture holds for hypersurfaces that are not ruled. Indeed, we may trivially bound $N_Y(B)$ as $O_{d,n}(B^{n-2})$ using Theorem 3.1 and the fact that Y has dimension at most $n-3$ when X is not ruled.

Let us now discuss how Theorem 3.3 can be used to deal with the dimension growth conjecture for small values of $n \geqslant 3$. In doing so we may suppose that X is ruled and is irreducible of degree at least 2. In particular the case $n = 3$ of curves is already dispatched, since the inequalities in (3.9) are vacuous in this case.

Turning to the case $n = 4$ of surfaces we may suppose that $\dim F_1(X) = 1$ by (3.9). Likewise, in view of the fact that X is irreducible and not a linear space, we have $F_2(X) = \emptyset$. It therefore follows from Theorem 3.3 and Lemma 3.1 that

$$\begin{aligned} N_X(B) &\ll_{\varepsilon,X} B^{2+\varepsilon} + \sum_{\substack{L\in F_1(X)(\mathbb{Q})\\ H(L)\ll_{\varepsilon,X} B^{1+\varepsilon}}} N_L(B) \\ &\ll_{\varepsilon,X} B^{2+\varepsilon} + \sum_{\substack{L\in F_1(X)(\mathbb{Q})\\ H(L)\ll_{\varepsilon,X} B^{1+\varepsilon}}} \left(1 + \frac{B^2}{H(L)}\right), \end{aligned}$$

where the first term is just the contribution from the linear spaces of dimension 0. We now break the summation over L into dyadic intervals according to the height of L. This gives

$$\begin{aligned} N_X(B) &\ll_{\varepsilon,X} B^{2+\varepsilon} + \sum_{P=2^j\ll_{\varepsilon,X} B^{1+\varepsilon}} \sum_{\substack{L\in F_1(X)(\mathbb{Q})\\ P\leqslant H(L)<2P}} \left(1 + \frac{B^2}{P}\right) \\ &\ll_{\varepsilon,X} B^{2+\varepsilon} + B^2 \sum_{P=2^j\ll_{\varepsilon,X} B^{1+\varepsilon}} \frac{1}{P} N_{F_1(X)}(2P). \end{aligned}$$

Now we have already seen that $F_1(X)$ has dimension 1. Suppose that one of its irreducible components is a line, Λ, say. Then the space of lines parametrised by Λ would cut out a plane contained in X, contradicting the fact that $F_2(X)$ is empty. Hence we may suppose that all of the irreducible components of $F_1(X)$, of which there are finitely many, have degree at least 2. An application of the dimension growth conjecture for curves therefore yields $N_{F_1(X)}(2P) = O_{\varepsilon,X}(P^{1+\varepsilon})$, giving

$$N_X(B) \ll_{\varepsilon,X} B^{2+\varepsilon} + B^2 \sum_{P=2^j\ll_{\varepsilon,X} B^{1+\varepsilon}} P^{\varepsilon} \ll_{\varepsilon,X} B^{2+2\varepsilon}.$$

This completes the proof of Conjecture 3.4 in the case $n = 4$, on redefining the choice of $\varepsilon > 0$.

The case $n = 5$ of threefolds is distinctly trickier. Here we would like to show that

$$N_X(B) \ll_{\varepsilon,X} B^{3+\varepsilon} \tag{3.10}$$

for irreducible threefolds $X \subset \mathbb{P}^4$. The main difficulty is that singular threefolds can contain a large number of lines and planes. Note that it follows from (3.9) that, in the case of interest, the dimension of $F_1(X)$ is either 2 or 3. When X is singular, there is an old result of Segre [110] which shows that $\dim F_1(X) = 3$ only when X is either a cone or a scroll of planes. In the latter case, when X is cubic, one can go slightly further and show that it must take the shape

$$x_1x_2x_3 + x_1^2x_4 + x_2^2x_5 = 0,$$

over $\overline{\mathbb{Q}}$. This is achieved in [30, Lemma 14] by considering generic hyperplane sections of X. It is easy to see that this variety contains the family of planes

$$\lambda x_1 - \mu x_2 = \lambda\mu x_3 + \mu^2 x_4 + \lambda^2 x_5 = 0,$$

for $[\lambda, \mu] \in \mathbb{P}^1$, in addition to the isolated plane $x_1 = x_2 = 0$. In particular it certainly contains a three-dimensional family of lines. A major obstacle in the proof of (3.10) is that each rational plane already contributes $\gg_X B^3$ to $N_X(B)$.

We will content ourselves with proving (3.10) for non-singular threefolds of degree at most 4. Then $F_2(X)$ is finite-dimensional and Lemma 3.5 implies that $F_1(X)$ has dimension at most 2. We argue now as in the treatment of the case $n = 4$ above. Thus an application of Theorem 3.3 and Lemma 3.1 gives

$$N_X(B) \ll_{\varepsilon,X} B^{3+\varepsilon} + \sum_{\substack{L \in F_1(X)(\mathbb{Q}) \\ H(L) \ll_{\varepsilon,X} B^{1+\varepsilon}}} \Big(1 + \frac{B^2}{H(L)}\Big),$$

where the first term is just the contribution from the linear spaces of dimension 0 or 2. We may clearly replace the summation over $L \in F_1(X)$ with a summation over lines belonging to irreducible components of $F_1(X)$ which have degree at least 2. Indeed, any lines parametrised by a linear component of $F_1(X)$ will produce points lying on a plane contained in X, and we have already remarked that there are finitely many of these, each one contributing $O_X(B^3)$ to $N_X(B)$. Thus, on breaking the summation into dyadic intervals and applying the dimension growth conjecture for surfaces, it follows that

$$N_X(B) \ll_{\varepsilon,X} B^{3+\varepsilon} + B^2 \sum_{P = 2^j \ll_{\varepsilon,X} B^{1+\varepsilon}} \frac{1}{P} \cdot P^{2+\varepsilon} \ll_{\varepsilon,X} B^{3+2\varepsilon}.$$

This therefore completes the proof of (3.10) in the case $d \in \{3, 4\}$ and $n = 5$, assuming that X is non-singular. Note that the case $d \geqslant 5$ and $n = 5$, with X non-singular, is trivially handled since such varieties are not ruled.

We end this section with the remark that when singularities are permitted, we have not yet been able to make these arguments work in the case $n = 6$ of cubic fourfolds. Here the problem is compounded by the fact that the Fano variety of lines can have dimension as large as 5.

Exercises for Chapter 3

Exercise 3.1. Adapt the proof of Lemma 3.2 to establish Lemma 3.3.

Exercise 3.2. Let $m \geqslant 1$ and let $X_{d,m}$ be the Fermat hypersurface

$$x_1^d + \cdots + x_{2m+2}^d = 0$$

in $\mathbb{P}^{2m+1}$. Show that $X_{d,m}$ contains precisely $c_m d^{m+1}$ m-planes, where

$$c_m = (2m+1)(2m-1)\cdots 3 \cdot 1.$$

Exercise 3.3. Adapt the proof of Theorem 3.1 to establish Lemma 3.6.

Exercise 3.4. By combining elimination theory with the sharper Lang–Weil bound in (3.4), show how the second estimate in Theorem 3.2 can be improved to

$$N_X(B) \ll_{d,n} B^{n-1-\frac{3}{2n-3}+\varepsilon},$$

for any $\varepsilon > 0$, assuming that X is irreducible and has degree at least 2.

Chapter 4

Uniform bounds for curves and surfaces

In 1989 Bombieri and Pila [7] pioneered a ground-breaking method for getting good upper bounds for the number of integer points on affine algebraic curves which are restricted to lie in a box. The most important feature of their estimate is its uniformity with respect to the particular curve. Thus if $C \subset \mathbb{A}^2$ is an irreducible plane curve of degree d with integer coefficients, then they establish a bound of the form

$$\#\big(C(\mathbb{Z}) \cap [-B, B]^2\big) = O_{d,\varepsilon}(B^{\frac{1}{d}+\varepsilon}), \tag{4.1}$$

for any $\varepsilon > 0$. Here the implied constant depends at most upon the choice of ε and the degree d. This is an extremely versatile result, which has already had a significant impact on many problems in analytic number theory. It also lies at the heart of an inductive proof (based on hyperplane sections) of the bound due to Pila [100], which appears as entry (iv) in Table 3.1. One notes that the Bombieri–Pila bound is essentially best possible, as consideration of the curve $x^d = y$ shows. One has $\gg B^{\frac{1}{d}}$ solutions of the form $x = a$, $y = a^d$ with modulus at most B.

The arguments of Bombieri and Pila make crucial use of properties of $\mathbb{A}^2$ and left open the question of whether there might be an extension of the method to higher dimensions and, indeed, to projective space. This has now been answered in the affirmative by Heath-Brown [74], as summarised in the following result.

Theorem 4.1 ("Theorem 14"). *Let $F \in \mathbb{Z}[x_1, \ldots, x_n]$ be an absolutely irreducible form of degree d, defining a hypersurface $X \subset \mathbb{P}^{n-1}$. Let $\varepsilon > 0$ and $B \geqslant 1$ be given. Then we can find $D, k \in \mathbb{N}$ such that the following hold:*

(i) *we have $D \ll_{d,n,\varepsilon} 1$ and $k \ll_{d,n,\varepsilon} B^{(n-1)d^{-\frac{1}{n-2}}+\varepsilon}(\log \|F\|)^{2n-3}$; and*

(ii) *there exist forms $F_1, \ldots, F_k \in \mathbb{Z}[x_1, \ldots, x_n]$, coprime to F with degrees at most D, such that every point counted by $N_X(B)$ lies on one of the hypersurfaces $F_j = 0$.*

Earlier we saw in Theorem 3.3 how we can cover all of the rational points of bounded height on X by a small number of linear spaces. The essence of Theorem 4.1 is that we can cover these points with an even smaller number of hypersurfaces. In both results the dependence of the estimates upon the coefficients of the underlying form F is made completely explicit, with all implied constants depending at most upon the degree of F and the number of variables involved. As with Theorem 3.3, it is with relatively small values of n that the greatest use can be made of Theorem 4.1.

The proof of Theorem 4.1 is based on a far-reaching generalisation of the *determinant method* that lies at the basis of [7]. We have decided not to give a comprehensive account of the method here, largely because such an overview is already served by the lecture notes compiled by Heath-Brown [77] for the C.I.M.E. *Summer School on Analytic Number Theory* in Cetraro in 2002. Nonetheless we will present a broad outline of the method in Section 4.1, below.

The dependence on the coefficients of the form F in Theorem 4.1 is mild but annoying. The following elementary result, due to Heath-Brown [74, Theorem 4], considers the possibility that the coefficients of F are very large compared to B. Recall that a form with integer coefficients is said to be *primitive* if the greatest common divisor of its coefficients is 1.

Lemma 4.1. *Let $F \in \mathbb{Z}[x_1, \dots, x_n]$ be a primitive non-zero form of degree $d \geqslant 2$, defining a hypersurface $X \subset \mathbb{P}^{n-1}$. Then either*

$$\|F\| \ll_{d,n} B^{d\theta}, \quad \theta = \binom{d+n-1}{n-1},$$

or else there exists a form $G \in \mathbb{Z}[x_1, \dots, x_n]$ of degree d, not proportional to F, such that G vanishes at each point counted by $N_X(B)$.

Proof. Let $\mathbf{x}^{(1)}, \dots, \mathbf{x}^{(N)} \in \mathbb{Z}^n_{\mathrm{prim}}$ be the complete set of representative vectors, corresponding to points on the hypersurface $F = 0$ with height at most B, that are counted by $N_X(B)$. Consider the $N \times \theta$ matrix $\mathbf{M}$ whose ith row consists of the θ possible monomials of degree d in the variables $x_1^{(i)}, \dots, x_n^{(i)}$. If the vector $\mathbf{F} \in \mathbb{Z}^\theta_{\mathrm{prim}}$ has entries which are the corresponding coefficients of F, we will have $\mathbf{MF} = \mathbf{0}$. Since $\mathbf{F} \neq \mathbf{0}$, it follows that $\mathbf{M}$ has rank at most $\theta - 1$, whence $\mathbf{MG} = \mathbf{0}$ has at most $O_{d,n}(1)$ non-zero integer solutions $\mathbf{G}$ constructed from the subdeterminants of $\mathbf{M}$, with $|\mathbf{G}| \ll_{d,m} B^{d\theta}$.

Let G be the form of degree d corresponding to the vector $\mathbf{G}$. Then G and F share N common zeros $\mathbf{x}^{(1)}, \dots, \mathbf{x}^{(N)}$. Either G is a rational multiple of F, in which case $\|F\| \ll_{d,n} B^{d\theta}$, or else G is not a rational multiple of F and the points $\mathbf{x}^{(i)}$ are all common solutions of the pair of equations $F(\mathbf{x}) = G(\mathbf{x}) = 0$. This therefore completes the proof of the lemma. □

Let us now consider the full effect of Theorem 4.1 for small values of n. Beginning with curves we have the following consequence ([74, Theorem 5]).

Theorem 4.2. *Let $\varepsilon > 0$ and let $C \subset \mathbb{P}^{n-1}$ be an irreducible curve of degree $d \geqslant 2$, defined over $\mathbb{Z}$. Then we have*

$$N_C(B) = O_{d,n,\varepsilon}(B^{\frac{2}{d}+\varepsilon}).$$

Proof. Suppose first that $C \subset \mathbb{P}^2$ is contained in the projective plane and is defined by a ternary form $F \in \mathbb{Z}[x_1, x_2, x_3]$. Taking $n = 3$, Theorem 4.1 shows that every point counted by $N_C(B)$ satisfies $F = F_j = 0$ for some $j \ll_{d,\varepsilon} B^{\frac{2}{d}+\varepsilon}(\log \|F\|)^3$. Bézout's Theorem shows that there are at most dD points for each j, so that

$$N_C(B) \ll_{d,\varepsilon} B^{\frac{2}{d}+\varepsilon}(\log \|F\|)^3.$$

The statement of Theorem 4.2 for plane curves then follows via an application of Lemma 4.1.

To deal with curves in arbitrary ambient projective space one needs an argument involving birational projections. Let $C \subset \mathbb{P}^{n-1}$ be an irreducible curve of degree $d \geqslant 2$. We need to construct a projection $C \to \mathbb{P}^2$, such that the image $\overline{C} \subset \mathbb{P}^2$ is an irreducible plane curve of degree d. Moreover we shall want to choose our projection in such a way that we have

$$N_C(B) \leqslant dN_{\overline{C}}(kB), \tag{4.2}$$

for some $k \ll_{d,n} 1$. Once this is accomplished, the statement of the theorem easily follows from our earlier treatment of plane curves.

To establish the existence of a suitable projection, we must first deal with the possibility that C is *degenerate*, by which we mean that $C \subset H$ for some hyperplane $H \in \mathbb{P}^{n-1*}$. Assume that H is defined by the linear equation $\mathbf{a}.\mathbf{x} = 0$ and suppose without loss of generality that $a_n \neq 0$. Then the point $y = [0, \ldots, 0, 1]$ is not contained in H and so the projection π_y from C onto the hyperplane $x_n = 0$ is a regular map. Moreover π_y is clearly birational onto the image $\overline{C} = \pi_y(C) \subset \mathbb{P}^{n-2}$, whence $\overline{C}$ is an irreducible variety of degree d such that $N_C(B) \leqslant N_{\overline{C}}(B)$. Now either $\overline{C}$ is not degenerate, or we may repeat the argument once again. It will therefore suffice to assume that C is not degenerate.

We wish to find a $(n-4)$-plane $\Lambda \subset \mathbb{P}^{n-1}$ such that the projection

$$\pi_\Lambda : C \to \Gamma$$

from Λ onto any plane Γ disjoint from Λ, is birational onto the image. Moreover, we will need Λ to be defined over $\mathbb{Q}$ and have height $H(\Lambda) \ll_{d,n} 1$ under the standard Plücker embedding. This is the content of [36, Lemmas 3 and 6], which we will not prove here. Let

$$Y = \{\Lambda \in \mathbb{G}(n-4, n-1) : \dim S_{\Lambda,C} \geqslant 1\},$$

where $S_{\Lambda,C}$ denotes the set of $M \in \mathbb{G}(n-3, n-1)$ such that $\Lambda \subset M$ and M intersects C in at least two (possibly coincident) points. Then [36, Lemma 6] implies that Y is a proper subvariety of $\mathbb{G}(n-4,n-1)$ of degree $O_{d,n}(1)$. Moreover, if Y' denotes the set of $\Lambda \in \mathbb{G}(n-4,n-1)$ which meet C, then Y' is also a proper subvariety and has degree $O_{d,n}(1)$. It follows from [36, Lemma 3] that there exists a $(n-4)$-plane $\Lambda \notin Y \cup Y'$, such that Λ is defined over $\mathbb{Q}$ and has height $H(\Lambda) \ll_{d,n} 1$. In particular $\Lambda \cap C$ is empty and the projection from Λ is birational onto its image.

Now suppose that Λ is equal to the linear span of $n-3$ points

$$[\mathbf{h}_1], \dots, [\mathbf{h}_{n-3}] \in \mathbb{P}^{n-1},$$

for vectors $\mathbf{h}_1, \dots, \mathbf{h}_{n-3} \in \mathbb{Z}^n$ of modulus $O_{d,n}(1)$. Then by [36, Lemma 6] we can select a vector $\mathbf{g}_1 \in \mathbb{Z}^n$ of modulus $O_{d,n}(1)$ such that $\mathbf{g}_1.\mathbf{h}_j = 0$ for $2 \leqslant j \leqslant n-3$ and $\mathbf{g}_1.\mathbf{h}_1 \neq 0$. Similarly, for $2 \leqslant i \leqslant n-3$ we may find vectors $\mathbf{g}_i \in \mathbb{Z}^n$ of modulus $O_{d,n}(1)$ such that $\mathbf{g}_i.\mathbf{h}_j = 0$ for $i \neq j$ and $\mathbf{g}_i.\mathbf{h}_i \neq 0$. These vectors are clearly linearly independent and define the plane

$$\Gamma: \quad \mathbf{g}_1.\mathbf{x} = \cdots = \mathbf{g}_{n-3}.\mathbf{x} = 0.$$

By construction $\Lambda \cap \Gamma$ is empty. We may now explicitly write our projection in the form

$$\pi_\Lambda(x) = \left[\mathbf{x} - \frac{(\mathbf{g}_1.\mathbf{x})}{(\mathbf{g}_1.\mathbf{h}_1)}\mathbf{h}_1 - \cdots - \frac{(\mathbf{g}_{n-3}.\mathbf{x})}{(\mathbf{g}_{n-3}.\mathbf{h}_{n-3})}\mathbf{h}_{n-3}\right],$$

for any $x = [\mathbf{x}] \in C$. This map is regular and birational onto its image. Moreover, the image $\overline{C} = \pi_\Lambda(C) \subset \Gamma$ is an irreducible curve of degree d and the fibre over any point of $\overline{C}$ contains at most d points.

Let $\lambda = \prod_i \mathbf{g}_i.\mathbf{h}_i$ and let $\lambda_j = \prod_{i \neq j} \mathbf{g}_i.\mathbf{h}_i$. Then since $|\mathbf{g}_i.\mathbf{x}| \leqslant n|\mathbf{g}_i||\mathbf{x}|$ for $1 \leqslant i \leqslant n-3$, we deduce that $H(\pi_\Lambda(x)) \leqslant kH(x)$ with

$$k = |\lambda| + n\sum_{i=1}^{n-3} |\lambda_i| H([\mathbf{g}_i]) H([\mathbf{h}_i]) \ll_{d,n} 1.$$

It therefore follows that (4.2) holds, as claimed. This completes the proof of Theorem 4.2. □

Later, in Lemma 7.4, we will meet a version of this result in which the representative coordinates for the points are restricted to lie in lopsided boxes. We have already seen in (2.4) that the exponent $\frac{2}{d}$ is best possible for rational curves, but not for curves of higher genus. In this context there is work by Ellenberg and Venkatesh [57] that merits mentioning. This establishes the existence of a real number $\delta > 0$ such that

$$N_C(B) = O_{d,n}(B^{\frac{2}{d}-\delta}),$$

for any irreducible curve $C \subset \mathbb{P}^{n-1}$ which has geometric genus $g \geqslant 1$. Of course, as mentioned in our discussion of curves in Section 2.2, much better estimates are available for individual curves. The salient feature here is that the estimate of Ellenberg–Venkatesh is completely uniform in the coefficients of the polynomials defining the curve. The value of δ that emerges from the proof is rather small, however, with $\delta = \frac{1}{450}$ being permissible for plane non-singular cubics. During the conference *Diophantine equations* in Bonn in 2009, Salberger announced that $\delta = \frac{2}{327}$ is permissible if $C(\mathbb{R})$ is connected. Under suitable hypotheses on the growth of the rank of elliptic curves Heath-Brown [73, Theorem 3] has shown that one can do much better than this.

Theorem 4.3. *Let $F \in \mathbb{Z}[x_1, x_2, x_3]$ be a non-singular cubic form defining an elliptic curve E in $\mathbb{P}^2$. Let $\varepsilon > 0$ and assume that Conjecture 2.7 holds. Then we have*

$$N_E(B) \ll_\varepsilon B^\varepsilon \|F\|^\varepsilon.$$

Here we recall the notation $\|F\|$ is used to denote the maximum modulus of the coefficients of the polynomial F. It is easy to extend the bound in Theorem 4.3 to other types of elliptic curves.

Theorem 4.4. *Let $F_1, F_2 \in \mathbb{Z}[x_1, \dots, x_4]$ be a pair of quadratic forms defining an elliptic curve E in $\mathbb{P}^3$. Let $\varepsilon > 0$ and assume that Conjecture 2.7 holds. Then we have*

$$N_E(B) \ll_\varepsilon B^\varepsilon \|F_1\|^\varepsilon \|F_1\|^\varepsilon.$$

Proof. Since E is elliptic it follows that $E(\mathbb{Q}) \neq \emptyset$. We may assume that E contains the rational point $[0, 0, 0, 1]$. But then the two quadratic forms take the shape

$$F_1(x_1, \dots, x_4) = x_4 L_1 + G_1, \quad F_2(x_1, \dots, x_4) = x_4 L_2 + G_2,$$

for forms $L_i, G_i \in \mathbb{Z}[x_1, x_2, x_3]$ of degrees 1 and 2, respectively. If L_1 and L_2 are linearly dependent then we may suppose that $L_2 = 0$. But then the intersection $F_1 = F_2 = 0$ defines a curve of genus 0, which is impossible. Hence L_1 and L_2 are linearly independent and we can eliminate x_4 to produce a cubic equation

$$L_1 G_2 - L_2 G_1 = 0.$$

The corresponding cubic curve contains the rational point given as the intersection point of the two lines $L_1 = 0$ and $L_2 = 0$.

In an entirely concrete fashion we have therefore produced a birational map θ which transforms E into a plane elliptic curve $E' \subset \mathbb{P}^2$. This map is obviously given by formulae in which the coefficients are rational functions involving the coefficients of F_1, F_2. Moreover, θ takes any rational point x counted by $N_E(B)$ to a rational point $y = y(x)$ on E', with height

$$H(y) \leqslant cB \|F_1\|^c \|F_2\|^c,$$

for a suitable absolute constant $c > 0$. Hence it follows from Theorem 4.3 that

$$N_E(B) \leqslant N_{E'}(cB \|F_1\|^c \|F_2\|^c) \ll_\varepsilon B^\varepsilon \|F_1\|^\varepsilon \|F_1\|^\varepsilon,$$

for any $\varepsilon > 0$. □

Let us now consider the case $n = 4$ of surfaces $X \subset \mathbb{P}^3$ in Theorem 4.1. One sees that the relevant points lie on $O_{d,\varepsilon}(B^{\frac{3}{\sqrt{d}}+\varepsilon}(\log\|F\|)^5)$ curves in X, each having degree at most dD. It is now possible to apply Theorem 4.2 to estimate the number of points on each such curve, as the following shows ([74, Theorem 7]).

Theorem 4.5. *Let $\varepsilon > 0$ and let $X \subset \mathbb{P}^3$ be an irreducible surface of degree d, defined over $\mathbb{Z}$. For each $e \leqslant d$, we let $U_e \subseteq X$ be the open subset formed by deleting all of the curves of degree $\leqslant e$ that are contained in X. Then we have*

$$N_{U_e}(B) = O_{d,\varepsilon}(B^{\frac{3}{\sqrt{d}}+\frac{2}{e+1}+\varepsilon}).$$

Proof. Since we are to ignore points lying on curves of degree $\leqslant e$ contained in X, it remains to consider the contribution from $O_{d,\varepsilon}(B^{\frac{3}{\sqrt{d}}+\varepsilon}(\log\|F\|)^5)$ irreducible curves in X which have degree $\delta \in \{e+1, \ldots, dD\}$. Such curves therefore contribute $O_{d,\varepsilon}(B^{\frac{2}{e+1}+\varepsilon})$ by Theorem 4.2. This implies that

$$N_{U_e}(B) = O_{d,\varepsilon}(B^{\frac{3}{\sqrt{d}}+\frac{2}{e+1}+2\varepsilon}(\log\|F\|)^5).$$

We complete the proof of the theorem by applying Lemma 4.1 and redefining the choice of $\varepsilon > 0$. □

In view of the fact $N_{\mathbb{P}^1}(B) \gg B^2$ we cannot in general hope for a version of this result with U_e replaced by X. When $e = 1$, Theorem 4.5 gives an upper bound for the number of points of height at most B in $X(\mathbb{Q})$, avoiding points on lines, which improves on the bound given by Conjecture 3.4 for surfaces of degree $d \geqslant 9$.

When X is non-singular we have much better control over the curves of low degree that can appear as divisors. The following result is due to Colliot-Thélène [74, Appendix].

Lemma 4.2. *Let $X \subset \mathbb{P}^3$ be a non-singular surface of degree $d \geqslant 3$. Then X contains $O_d(1)$ curves of degree $\leqslant d-2$.*

Applying Lemma 4.2 in the proof of Theorem 4.5 we are easily led to the following refinement ([74, Theorem 11]).

Theorem 4.6. *Let $\varepsilon > 0$ and let $X \subset \mathbb{P}^3$ be a non-singular surface of degree $d \geqslant 3$, defined over $\mathbb{Z}$. Let U_e be as in the statement of Theorem 4.5. Then we have*

$$N_{U_e}(B) = O_{d,\varepsilon}(B^{\frac{2}{e+1}+\varepsilon} + B^{\frac{3}{\sqrt{d}}+\frac{2}{\max\{d-1,e+1\}}+\varepsilon}).$$

The reader is invited to compare Theorem 4.6 with the 4th entry in Table 1.1. In fact most of the work in [21] is dedicated to handling separately the contribution from the conics in the surface (1.8), thereby paving the way towards an application of Theorem 4.6 with $e = 2$.

Since the work of Heath-Brown the determinant method has evolved quite considerably, to the extent that the bounds in Theorems 4.5 and 4.6 have been greatly refined. The main player in this evolution has been Salberger (see [103, 104, 105, 106]). One of the major breakthroughs achieved by Salberger is a means of slicing by hypersurfaces twice in one go. This is the outcome of [106]. In the setting of surfaces it gives the following result.

Theorem 4.7. *Let $\varepsilon > 0$ and let $X \subset \mathbb{P}^{n-1}$ be an irreducible surface of degree d, defined over $\mathbb{Z}$. Then there exists a set of $O_{d,n,\varepsilon}(B^{\frac{3}{2\sqrt{d}}+\varepsilon})$ curves of degree $O_{d,n}(1)$ on X such that there are at most $O_{d,n,\varepsilon}(B^{\frac{3}{\sqrt{d}}+\varepsilon})$ points counted by $N_X(B)$ which do not lie on any of these curves.*

Recall the definition of the open subset U_e from the statement of Theorem 4.5. It is now easy to invoke Theorem 4.7 in the proofs of Theorems 4.5 and 4.6 to get improvements. The following consequence deals with the former result alone.

Corollary 4.1. *Let $\varepsilon > 0$ and let $X \subset \mathbb{P}^3$ be an irreducible surface of degree d, defined over $\mathbb{Z}$. Then we have*

$$N_{U_e}(B) = O_{d,\varepsilon}(B^{\frac{3}{\sqrt{d}}+\varepsilon} + B^{\frac{3}{2\sqrt{d}}+\frac{2}{e+1}+\varepsilon}).$$

In (2.1) we met the definition of the quantity β_V associated to a projective algebraic variety $V \subset \mathbb{P}^{n-1}$. So far we have seen many examples for which β_V is expected to be integer, at least conjecturally. This is certainly the case in all successful applications of the Hardy–Littlewood circle method, for example. It would be peculiar to imagine that the following conjecture is false.

Conjecture 4.1. *We have $\beta_V \in \mathbb{Q}$ for any variety $V \subset \mathbb{P}^{n-1}$.*

We proceed to consider some situations for which β_V is fractional, by which we mean $\beta_V \in \mathbb{Q} \setminus \mathbb{Z}$. This will build on some of the results that we have already met. Firstly, we recall from (2.4) that $\beta_V = \frac{2}{d}$ for irreducible curves $V \subset \mathbb{P}^{n-1}$ of degree d and genus 0, provided that $V(\mathbb{Q}) \neq \emptyset$. Thus we get fractional values of β_V as soon as $d \geqslant 3$. Let us consider the situation for irreducible surfaces $V \subset \mathbb{P}^3$ of degree d. For the case $d = 3$ we have seen that β_V is expected to be 1 when V contains no lines defined over $\mathbb{Q}$, but that it should take the value 2 in every other case. When $d \geqslant 4$ and V is non-singular Conjecture 2.1 implies that β_V ought to be zero. When one allows singularities one can quite easily construct examples which lead to fractional values of β_V.

In degree 4 there is the *Steiner surface*

$$V: \quad x_1^2x_2^2 - x_1^2x_3^2 + x_2^2x_3^2 - x_1x_2x_3x_4 = 0.$$

All of the rational points $x = [\mathbf{x}] \in V(\mathbb{Q})$ are parametrised via

$$\mathbf{x} = (u_2u_3, u_1u_3, u_1u_2, u_1^2 + u_2^2 + u_3^2),$$

for vectors $\mathbf{u} = (u_1, u_2, u_3) \in \mathbb{Z}^3_{\text{prim}}$. One can use this to show that $\beta_V = \frac{3}{2}$ for this example. Now let

$$V_d: \quad x_1^d + x_2^d - x_3^{d-1}x_4 = 0. \tag{4.3}$$

It is easy to check that $V_d \subset \mathbb{P}^3$ has finite singular locus. In particular it is irreducible. By Exercise 4.1 we have

$$N_{V_d}(B) \gg_d B^{\frac{3}{d}}.$$

The upper bound problem is more challenging.

We begin with the observation that the lines contained in V_d are given by

$$L_\omega^{(i)}: \quad x_1 = \omega x_2, \; x_i = 0,$$

for $i \in \{3,4\}$ and ω a dth root of -1. Thus when d is even, we obtain a contribution of $O(1)$ to $N_{V_d}(B)$ from points lying on lines contained in the surface. Turning to curves of higher degree, although we will not include details here, one can draw on the work of Browning [21] to show that the surface contains no quadrics which are irreducible over $\mathbb{Q}$ when $d \geqslant 6$ is even. The underlying idea is to show that apart from $O_d(1)$ exceptions, which correspond to the family of lines $L_\omega^{(3)}, L_\omega^{(4)}$, any choice of non-zero $\mathbf{y} \in \overline{\mathbb{Q}}^4$ leads to a $\mathbb{Q}$-irreducible plane section

$$C_{\mathbf{y}}: \quad \mathbf{x}.\mathbf{y} = x_1^d + x_2^d - x_3^{d-1}x_4 = 0.$$

For this one makes a detailed analysis of the possible singularities taken by $C_{\mathbf{y}}$. Taking this work on faith it is now clear that $N_{V_d}(B) = N_{U_2}(B) + O_d(1)$, where $U_2 \subset V_d$ is the open subset formed by deleting all of the curves of degree at most 2 that are contained in the surface V_d. The upper bound

$$N_{V_d}(B) \ll_{d,\varepsilon} B^{\frac{3}{2\sqrt{d}}+\frac{2}{3}+\varepsilon}$$

now follows from Corollary 4.1, for even $d \geqslant 6$.

Combining the lower and upper bounds for $N_{V_d}(B)$, we clearly have

$$0 < \frac{3}{d} < \frac{3}{2\sqrt{d}} + \frac{2}{3} < 1$$

for $d > 20$. It follows that β_{V_d} is fractional for every even degree $d \geqslant 22$.

4.1 The determinant method

In this section we discuss the estimate (4.1) and its subsequent generalisation leading to Theorem 4.1. Loosely speaking, the first step in the approach adopted by Bombieri and Pila [7] is to divide the affine curve $C \subset \mathbb{A}^2$ into many small patches, where "small" here means small in the sense of the real Euclidean metric. The number of patches is $O_{d,\varepsilon}(B^{\frac{1}{d}+\varepsilon})$, where d is the degree of C. For the integer points with modulus $\leqslant B$ in each patch one then finds an auxiliary curve of degree $O_d(1)$ which meets each of these points. In total, therefore, one finds $O_{d,\varepsilon}(B^{\frac{1}{d}+\varepsilon})$ auxiliary curves of degree $O_d(1)$ which cover all of the points in $C(\mathbb{Z})$ of height $\leqslant B$. An application of Bézout's theorem then yields the required bound.

The approach taken by Heath-Brown to establish Theorem 4.1 is similar in spirit to this, but rather than working with the real metric in the division of the variety into patches, he works with the p-adic metric for an auxiliary prime p. One needs to analyse the sets

$$S(F;B,p) := \{\mathbf{x} \in \mathbb{Z}^n : F(\mathbf{x}) = 0, \ |\mathbf{x}| \leqslant B, \ p \nmid \nabla F(\mathbf{x})\}$$

and

$$S(F;B) := \{\mathbf{x} \in \mathbb{Z}^n : F(\mathbf{x}) = 0, \ |\mathbf{x}| \leqslant B, \ \nabla F(\mathbf{x}) \neq \mathbf{0}\}.$$

The points $\mathbf{x} \in \mathbb{Z}^n$ for which $F(\mathbf{x}) = 0$, $|\mathbf{x}| \leqslant B$ and $\nabla F(\mathbf{x}) = \mathbf{0}$ are satisfactory from the point of view of Theorem 4.1, since one can then take $\frac{\partial F}{\partial x_i}$ as one of the auxiliary forms, for some $1 \leqslant i \leqslant n$. Assuming that $r = [\log \|F\|B]$ and $P \geqslant (\log \|F\|B)^2$, it is then possible (see [74, Lemma 4]) to find primes $p_1 < \cdots < p_r$ in the range $P \ll_{d,n} p_i \ll_{d,n} P$, such that

$$S(F;B) = \bigcup_{i=1}^{r} S(F;B,p_i).$$

This is the only place in the argument where a dependence on $\|F\|$ occurs.

Let $p = p_i$ for some $1 \leqslant i \leqslant r$. The next step is to break the points in $S(F;B,p)$ into projective equivalence classes

$$S(F;B,p) = \bigcup_{\mathbf{t}} S(\mathbf{t}),$$

where

$$S(\mathbf{t}) := \{\mathbf{x} \in S(F;B,p) : \mathbf{x} \equiv \rho\mathbf{t} \pmod{p} \text{ for some } \rho \in \mathbb{Z}\},$$

and the union is over all non-singular solutions $\mathbf{t} \in \mathbb{F}_p^n$ of $F(\mathbf{t}) \equiv 0 \pmod{p}$, up to projective equivalence. The kernel of the proof of Theorem 4.1 involves showing that for sufficiently large p, compared with B, all points $\mathbf{x} \in S(\mathbf{t})$ satisfy an auxiliary equation $F(\mathbf{x};\mathbf{t}) = 0$. The forms $F(\mathbf{x};\mathbf{t})$ will have the properties described in Theorem 4.1, so that we may take

$$k \ll_{d,n} \sum_{1 \leqslant i \leqslant r} p_i^{n-2} \ll_{d,n} P^{n-2}(\log \|F\|B),$$

by Lemma 3.6.

Clearly there is no loss of generality in assuming that the monomial x_n^d has non-zero coefficient in $F(\mathbf{x})$. For a given large integer D define the set

$$\mathscr{E} := \{\mathbf{e} \in \mathbb{Z}_{\geqslant 0}^n : e_n < d, \ e_1 + \cdots + e_n = D\}.$$

Let

$$E = \#\mathscr{E} = \binom{D+n-1}{n-1} - \binom{D-d+n-1}{n-1}.$$

Suppose for the moment that $E \leqslant \#S(\mathbf{t})$ and let $\mathbf{x}^{(1)}, \ldots, \mathbf{x}^{(E)}$ be distinct vectors in $S(\mathbf{t})$. At the heart of Heath-Brown's proof is an analysis of the $E \times E$ determinant

$$\Delta := \det(\mathbf{x}^{(i)\mathbf{e}})_{\substack{1 \leqslant i \leqslant E, \\ \mathbf{e} \in \mathscr{E}}}$$

where one writes $\mathbf{w}^{\mathbf{e}} := w_1^{e_1} \cdots w_n^{e_n}$. Thus Δ has rows corresponding to the different vectors $\mathbf{x}^{(i)}$ and columns corresponding to the various exponent n-tuples $\mathbf{e}$. The primary objective is to show that this determinant vanishes when p is sufficiently large.

On the one hand, Heath-Brown employs the implicit function theorem to show that for the $\mathbf{x}^{(i)\mathbf{e}}$ appearing in the determinant, there exists a positive integer m such that $p^m \mid \Delta$. This is the hardest part of the argument. On the other hand, it is trivial to see that $|\Delta| \leqslant E^E B^{DE}$, since each entry in the matrix has modulus at most B^D. Hence, if one chooses P to be sufficiently large, the choice

$$P \gg_{d,n} B^{(\frac{n-1}{n-2})d^{-\frac{1}{n-2}}+\varepsilon}(\log \|F\|)^2$$

being expedient, then one can indeed deduce that $\Delta = 0$. We refer the reader to [74] or [77] for details.

Everything is now in place to construct the form F_j that corresponds to our chosen value of $\mathbf{t}$. Let $K = \#S(\mathbf{t})$ and consider the matrix

$$\mathbf{M} := (\mathbf{x}^{(i)\mathbf{e}})_{\substack{1 \leqslant i \leqslant K, \\ \mathbf{e} \in \mathscr{E}}}$$

where the vectors $\mathbf{x}^{(i)}$ now run over all elements of $S(\mathbf{t})$. If $K \leqslant E - 1$, then it trivially follows that this matrix has rank at most $E - 1$. Otherwise, if $K \geqslant E$, then the above discussion shows that every $E \times E$ minor vanishes, so that $\mathbf{M}$ has rank at most $E - 1$ in this case too. It therefore follows that $\mathbf{Mc} = \mathbf{0}$ for some non-zero vector $\mathbf{c} \in \mathbb{Z}^E$. Hence if

$$F_j(\mathbf{x}) = \sum_{\mathbf{e} \in \mathscr{E}} c_{\mathbf{e}} \mathbf{x}^{\mathbf{e}},$$

we will have a non-zero form, of degree D, which vanishes for every $\mathbf{x} \in S(\mathbf{t})$. Moreover, it is clear from our choice of exponent set $\mathscr{E}$ that F and F_j are coprime since F contains a term in x_n^d and F_j does not.

4.2 The geometry of numbers

We will be concerned here with *lattices* $\mathsf{\Lambda} \subseteq \mathbb{Z}^n \subset \mathbb{R}^n$, which are just $\mathbb{Z}$-submodules of $\mathbb{Z}^n$, or equivalently, discrete subgroups of $\mathbb{Z}^n$ containing the origin $\mathbf{0}$. Our discussion is not intended as a comprehensive account of lattices and the geometry of numbers, for which the reader is recommended to consult the book by Cassels

[40]. Throughout this section we will write $\|\mathbf{x}\| := \sqrt{x_1^2 + \cdots + x_n^2}$ for the usual Euclidean norm on $\mathbb{R}^n$.

We say that a lattice Λ is *primitive* if it has a basis $\mathbf{b}_1, \ldots, \mathbf{b}_r$ that can be extended to a basis of $\mathbb{Z}^n$. If there is one basis which can be so extended, then any basis has such an extension. A necessary and sufficient condition for Λ to be primitive is that it should not be properly contained in any other r-dimensional sublattice of $\mathbb{Z}^n$. Henceforth, whenever we talk of a lattice, we will mean a primitive lattice, unless we explicitly say otherwise.

If Λ has dimension r, and $\mathbf{b}_1, \ldots, \mathbf{b}_r$ is a basis for Λ, we define the *determinant* $\det \Lambda$ of Λ to be the r-dimensional volume of the parallelepiped generated by $\mathbf{b}_1, \ldots, \mathbf{b}_r$. This is independent of the choice of the basis $\mathbf{b}_1, \ldots, \mathbf{b}_r$, and it follows from Hadamard's inequality that

$$\det \Lambda \leqslant \prod_{i=1}^{r} \|\mathbf{b}_i\|. \tag{4.4}$$

Given an integer sublattice $\Lambda \subseteq \mathbb{Z}^n$ there always exists a choice of basis which we will refer to as *minimal*, and which is extremely useful in our work. The basic properties of a minimal basis are recorded in the following result, which is originally due to Davenport [49, Lemma 5].

Lemma 4.3. *Let $\Lambda \subseteq \mathbb{Z}^n$ be a lattice of dimension r. Then Λ has a basis $\mathbf{b}^{(1)}, \ldots, \mathbf{b}^{(r)}$ such that if one writes $\mathbf{x} \in \Lambda$ as $\mathbf{x} = \sum_{i=1}^{r} \lambda_i \mathbf{b}^{(i)}$, then*

$$\lambda_i \ll_n \frac{\|\mathbf{x}\|}{\|\mathbf{b}^{(i)}\|}, \quad (1 \leqslant i \leqslant r), \tag{4.5}$$

where the implied constant is independent of Λ. Moreover, one has

$$\det \Lambda \leqslant \prod_{i=1}^{r} \|\mathbf{b}^{(i)}\| \ll_n \det \Lambda, \tag{4.6}$$

and

$$\|\mathbf{b}^{(1)}\| \leqslant \cdots \leqslant \|\mathbf{b}^{(r)}\|. \tag{4.7}$$

Proof. Let $\mathbf{b}^{(1)}$ denote any non-zero element of Λ for which $\|\mathbf{b}^{(1)}\|$ is least. Next, let $\mathbf{b}^{(2)} \in \Lambda \setminus \langle \mathbf{b}^{(1)} \rangle_{\mathbb{Z}}$ such that $\|\mathbf{b}^{(2)}\|$ is least, where $\langle \mathbf{a}_1, \ldots, \mathbf{a}_m \rangle_{\mathbb{Z}}$ denotes the $\mathbb{Z}$-linear span of vectors $\mathbf{a}_1, \ldots, \mathbf{a}_m \in \mathbb{Z}^n$. Choosing

$$\mathbf{b}^{(3)} \in \Lambda \setminus \langle \mathbf{b}^{(1)}, \mathbf{b}^{(2)} \rangle_{\mathbb{Z}}$$

in the same way, and continuing, one ultimately obtains a basis $\mathbf{b}^{(1)}, \ldots, \mathbf{b}^{(r)}$ of Λ such that (4.7) holds. If we set $s_i = \|\mathbf{b}^{(i)}\|$, for $1 \leqslant i \leqslant r$, then $s_1, \ldots, s_r$ are actually the successive minima of Λ, with respect to the Euclidean norm. The upper bound in (4.6) therefore follows from [40, Theorem I of Section VIII.2], whereas the lower bound follows from (4.4).

Now let Λ_j denote the $(r-1)$-dimensional lattice generated by

$$\mathbf{b}^{(1)},\dots,\mathbf{b}^{(j-1)},\mathbf{b}^{(j+1)},\dots,\mathbf{b}^{(r)}.$$

The perpendicular distance of the point $\mathbf{b}^{(j)}$ from the $(r-1)$-dimensional vector space spanned by the other basis points is $(\det\Lambda_j)^{-1}(\det\Lambda)$. Given a point $\mathbf{x}=\lambda_1\mathbf{b}^{(1)}+\cdots+\lambda_r\mathbf{b}^{(r)}\in\Lambda$ for $\lambda_1,\dots,\lambda_r\in\mathbb{Z}$, we deduce that the perpendicular distance of $\mathbf{x}$ from this vector space is $|\lambda_j|(\det\Lambda_j)^{-1}(\det\Lambda)$. Hence $\|\mathbf{x}\|\gg_n|\lambda_j|(\det\Lambda_j)^{-1}(\det\Lambda)$. But

$$\det\Lambda_j\leqslant s_1\cdots s_{j-1}s_{j+1}\cdots s_r\ll_n\frac{\det\Lambda}{s_j},$$

by (4.4), whence (4.5) holds. This completes the proof of Lemma 4.3. □

We now turn to the notion of dual lattices. We define the *dual* of a primitive lattice $\Lambda\subseteq\mathbb{Z}^n$ to be the lattice

$$\Lambda^*:=\{\mathbf{x}\in\mathbb{Z}^n:\mathbf{x}.\mathbf{y}=0\ \forall\mathbf{y}\in\Lambda\}.$$

Λ^* will always be primitive, with dimension $n-r$. Moreover if Λ is primitive, then $(\Lambda^*)^*=\Lambda$ and

$$\det\Lambda^*=\det\Lambda.$$

These facts are all established in Heath-Brown [68, Section 2], the final assertion being [68, Lemma 1].

Lemma 4.4. *Let* $\mathbf{a}\in\mathbb{Z}^n_{\mathrm{prim}}$. *The set*

$$\Lambda_{\mathbf{a}}:=\{\mathbf{x}\in\mathbb{Z}^n:\mathbf{a}.\mathbf{x}=0\}$$

is a lattice of dimension $n-1$ *and determinant* $\det\Lambda=\|\mathbf{a}\|$.

Proof. To establish the lemma we merely note that $\Lambda_{\mathbf{a}}$ is the dual of $\langle\mathbf{a}\rangle_{\mathbb{Z}}$, which is a lattice of dimension 1 and determinant $\|\mathbf{a}\|$. □

We will also need some information about the density of points on lattices that are confined to certain regions. Recall the notation $|\cdot|$ for the norm $|\mathbf{x}|=\max_{1\leqslant i\leqslant n}|x_i|$.

Lemma 4.5. *Let* $\Lambda\subseteq\mathbb{Z}^n$ *be a lattice of dimension* r. *Then we have*

$$\#\{\mathbf{x}\in\Lambda:|\mathbf{x}|\leqslant R\}\ll_n R^{r-1}+\frac{R^r}{\det\Lambda},$$

for any $R\geqslant 1$. *When* $r=2$ *we have*

$$\#\{\mathbf{x}\in\Lambda:|\mathbf{x}|\leqslant R,\ \gcd(x_1,\dots,x_n)=1\}\ll_n 1+\frac{R^2}{\det\Lambda}.$$

Proof. We apply Lemma 4.3. This shows that the number of vectors to be counted is at most the number of vectors $(\lambda_1,\ldots,\lambda_r)\in\mathbb{Z}^r$ such that $\lambda_i \ll_n \|\mathbf{b}^{(i)}\|^{-1}R$ for $1\leqslant i\leqslant r$. Since $\|\mathbf{b}^{(r)}\|\geqslant\cdots\geqslant\|\mathbf{b}^{(1)}\|\geqslant 1$, this gives us a bound

$$\ll_n \prod_{i=1}^{r}\Big(1+\frac{R}{\|\mathbf{b}^{(i)}\|}\Big)\ll_n R^{r-1}+\frac{R^r}{\prod_i\|\mathbf{b}^{(i)}\|}\ll_n R^{r-1}+\frac{R^r}{\det\mathsf{\Lambda}},$$

as required for the first part of the lemma. To see the second part, we note that there are $O_n(R^2(\det\mathsf{\Lambda})^{-1})$ relevant vectors if $\|\mathbf{b}^{(r)}\|\ll_n R$, whereas the alternative hypothesis contributes only $O_n(1)$, since the only primitive vectors $(\lambda_1,\lambda_2)\in\mathbb{Z}^2$ with $\lambda_2=0$ have $\lambda_1=\pm1$. ☐

This result essentially corresponds to counting rational points of bounded height on projective linear spaces, providing the promised proof of Lemma 3.1. Indeed, if $M\in\mathbb{G}(m,N)(\mathbb{Q})$ is a linear space of dimension m defined over $\mathbb{Q}$, then

$$\mathsf{M}=\{\mathbf{x}\in\mathbb{Z}^{N+1}:[\mathbf{x}]\in M\}\cup\{\mathbf{0}\}$$

is a lattice of rank $m+1$ whose determinant $\det\mathsf{M}$ has the same order of magnitude as the height $H(M)$ of M.

Recall the definition of $\mathsf{\Lambda}_\mathbf{a}$ from Lemma 4.4. When $n=3$, it follows from Lemma 4.5 that

$$\#\{\mathbf{x}\in\mathsf{\Lambda}_\mathbf{a}:|\mathbf{x}|\leqslant R,\ \gcd(x_1,x_2,x_3)=1\}\ll 1+\frac{R^2}{|\mathbf{a}|}.$$

It will be useful to have a refinement of this in which points lying in general boxes are allowed. The following result is due to Heath-Brown [68, Lemma 3].

Lemma 4.6. *Let* $\mathbf{a}\in\mathbb{Z}^3_{\mathrm{prim}}$ *and let* $R_1,R_2,R_3>0$. *Then we have*

$$\#\{\mathbf{x}\in\mathsf{\Lambda}_\mathbf{a}:|x_i|\leqslant R_i,\ \gcd(x_1,x_2,x_3)=1\}\ll 1+\frac{R_1R_2R_3}{\max_i|a_i|R_i}.$$

Proof. We begin by noting that since $\mathbf{x}$ is confined to the box $|x_i|\leqslant R_i$, each vector we are interested in lies in the ellipsoid $\Sigma(\mathbf{x})\leqslant 1$, with underlying matrix $\mathbf{Q}=3^{-1}\,\mathrm{Diag}(R_1^{-2},R_2^{-2},R_3^{-2})$. The intersection of this ellipsoid with the plane $\mathbf{a}.\mathbf{x}=0$ produces a further ellipse $S\subset\mathbb{R}^2$, say, with area

$$A_S=\pi\|\mathbf{a}\|\|\mathbf{Q}^{-1}(\mathbf{a})\det\mathbf{Q}\|^{-\frac12}\ll|\mathbf{a}|\frac{R_1R_2R_3}{\max_i|a_i|R_i}.$$

We now carry out a unimodular transformation $\mathbf{T}$ of $\mathbb{R}^3$ which takes S to a circle $\mathbf{T}S$ of area A_S, and $\mathsf{\Lambda}_\mathbf{a}$ to a lattice $\mathbf{T}\mathsf{\Lambda}_\mathbf{a}=\{\mathbf{Tx}:\ \mathbf{x}\in\mathsf{\Lambda}_\mathbf{a}\}$ of determinant $\|\mathbf{a}\|$. We are now interested in counting vectors in $\mathbb{Z}^3_{\mathrm{prim}}$ that belong to a lattice of rank 2 and determinant $\|\mathbf{a}\|$, and which are constrained to lie in a circle of radius $R\ll A_S^{\frac12}$. An application of Lemma 4.5 therefore finishes the proof. ☐

4.3 General plane curves

In this section we will establish a proof of Theorem 3.3 in the case $n = 3$ of curves. In fact we will go slightly further and deal with the possibility that the points in which we are interested are permitted to lie in lopsided boxes.

Throughout this section we let $F \in \mathbb{Z}[x_1, x_2, x_3]$ be an absolutely irreducible form of degree $d \geqslant 2$, defining a plane curve $C \subset \mathbb{P}^2$. Given $B_1, B_2, B_3 \geqslant 1$, we will be interested in the counting function

$$N(F; \mathbf{B}) = \#\{\mathbf{x} \in \mathbb{Z}^3_{\text{prim}} : F(\mathbf{x}) = 0,\ |x_i| \leqslant B_i\ (1 \leqslant i \leqslant 3)\}. \tag{4.8}$$

We clearly get information about the usual counting function $N_C(B)$ by taking $B_i = B$ for $1 \leqslant i \leqslant 3$. As we saw in Section 3.4, the case $n = 3$ of Theorem 3.3 basically says that all of the points counted by $N_C(B)$ lie on one of at most $O_{d,\varepsilon}(B^{1+\varepsilon})$ linear spaces of dimension 0 contained in C. This follows from the fact that C cannot contain any linear spaces of dimension exceeding 0 under the present hypotheses. Thus, when $n = 3$, Theorem 3.3 basically amounts to the statement that $N_C(B) = O_{d,\varepsilon}(B^{1+\varepsilon})$. We will prove the sharper bound

$$N_C(B) \ll_d B, \tag{4.9}$$

for any irreducible plane curve $C \subset \mathbb{P}^2$ of degree $d \geqslant 2$. It is easy to see that this is best possible when $d = 2$, as follows from (2.4), for example.

In view of Theorem 4.2 it remains to give a satisfactory treatment of the case $d = 2$, since one gets an exponent $\frac{2}{d}+\varepsilon < 1$ for $d \geqslant 3$. Rather than establishing (4.9) directly, we will attempt instead to give a bound for the more general counting function $N(F; \mathbf{B})$, but we shall restrict ourselves to non-singular curves. This is satisfactory for (4.9) since a conic is irreducible if and only if it is non-singular. We will establish the following result.

Theorem 4.8. *Let $F \in \mathbb{Z}[x_1, x_2, x_3]$ be a non-singular form with degree $d \geqslant 2$. Then we have*

$$N(F; \mathbf{B}) \ll_d (B_1 B_2 B_3)^{\frac{1}{3}}.$$

In establishing Theorem 4.8 we may assume without loss of generality that F is primitive, since we can always remove any factor common to the coefficients in the equation $F = 0$. Let us write

$$V = B_1 B_2 B_3, \quad B = \max_{1 \leqslant i \leqslant 3} B_i$$

for short, and let Δ be the discriminant of F. Thus Δ is non-zero and it is a polynomial of degree $3(d-1)^2$ in the coefficients of F, defined over $\mathbb{Z}$. Let

$$r = \left[\frac{9}{2}(d-1)^2 d(d+1)(d+2)\right] + 1. \tag{4.10}$$

By Bertrand's postulate we can choose primes $p_1, \ldots, p_r$, with

$$cV^{\frac{1}{3}} \leqslant p_1 < \cdots < p_r \ll_d V^{\frac{1}{3}}, \tag{4.11}$$

where c is an absolute constant to be chosen in due course. Now either there exists $1 \leqslant i \leqslant r$ for which $p_i \nmid \Delta$, or else

$$|\Delta| \geqslant \prod_{i=1}^{r} p_i \geqslant c^r V^{\frac{r}{3}}.$$

We begin by disposing of the latter case. Since $|\Delta| \ll_d \|F\|^{3(d-1)^2}$ it will follow that

$$\|F\| \gg_d V^{\frac{r}{9(d-1)^2}} \geqslant B^{\frac{r}{9(d-1)^2}}.$$

We now apply Lemma 4.1 with $n = 3$, so that $\theta = \frac{1}{2}(d+1)(d+2)$. This shows that if r is chosen as in (4.10), then there exists a ternary form G of degree d, not proportional to F, such that every $\mathbf{x}$ counted by $N(F;\mathbf{B})$ also satisfies the equation $G(\mathbf{x}) = 0$. But then Bézout's theorem gives $N(F;\mathbf{B}) \leqslant d^2$, which is satisfactory.

We may now concentrate on the case in which there is a prime p in the range $cV^{\frac{1}{3}} \leqslant p \ll_d V^{\frac{1}{3}}$, with the property that $p \nmid \Delta$. Our argument depends on the following result, which is a special case of [30, Lemma 7].

Lemma 4.7. *Let $H \in \mathbb{Z}[x_1, x_2, x_3]$ be a form of degree d and let p be a prime. Let $\mathbf{x} \in \mathbb{Z}^3$ be a vector for which $p \mid H(\mathbf{x})$ and*

$$p \nmid \nabla H(\mathbf{x}). \tag{4.12}$$

Then there exists $\mathbf{x}^{(1)} \in \mathbb{Z}^3$ satisfying $p \nmid \nabla H(\mathbf{x}^{(1)})$ and $p^2 \mid H(\mathbf{x}^{(1)})$, with $\mathbf{x}^{(1)} \equiv \mathbf{x} \pmod{p}$. Write

$$\mathsf{M} = \{\mathbf{w} \in \mathbb{Z}^3 : \ \mathbf{w} \equiv \rho\mathbf{x} \pmod{p} \text{ for some } \rho \in \mathbb{Z} \text{ and } p^2 \mid \mathbf{w}.\nabla H(\mathbf{x}^{(1)})\}.$$

Then M is independent of the choice of $\mathbf{x}^{(1)}$. Moreover, M is a lattice of dimension 3 and determinant $\det \mathsf{M} = p^3$. Finally, if $\mathbf{t} \in \mathbb{Z}^3_{\text{prim}}$ satisfies $H(\mathbf{t}) = 0$, and if there exists $\lambda \in \mathbb{Z}$ for which $\mathbf{t} \equiv \lambda\mathbf{x} \pmod{p}$, then $\mathbf{t} \in \mathsf{M}$.

Proof. The existence of a suitable $\mathbf{x}^{(1)}$ is an immediate consequence of Hensel's lemma. Specifically, on writing $\mathbf{x}^{(1)} = \mathbf{x} + p\mathbf{y}^{(1)}$, we find that p^2 divides $H(\mathbf{x}^{(1)})$ if and only if $\mathbf{y}^{(1)}.\nabla H(\mathbf{x}) \equiv -p^{-1}H(\mathbf{x}) \pmod{p}$. This is always solvable for $\mathbf{y}^{(1)}$, by (4.12).

Now suppose that $\mathbf{x}^{(2)} = \mathbf{x} + p\mathbf{y}^{(2)}$, with $\mathbf{y}^{(2)}.\nabla H(\mathbf{x}) \equiv -p^{-1}H(\mathbf{x}) \pmod{p}$ and $\mathbf{w} \equiv \rho\mathbf{x} \pmod{p}$. To show that M is independent of the choice of $\mathbf{x}^{(1)}$ it will suffice to demonstrate that $p^2 \mid \mathbf{w}.\nabla H(\mathbf{x}^{(1)})$ if and only if $p^2 \mid \mathbf{w}.\nabla H(\mathbf{x}^{(2)})$.

However, assuming that H has degree d, we have

$$\begin{aligned}\mathbf{w}.\nabla H(\mathbf{x}^{(i)}) &= \mathbf{w}.\nabla H(\mathbf{x}+p\mathbf{y}^{(i)})\\ &\equiv \mathbf{w}.\nabla H(\mathbf{x}) + p\sum_{j,k=1}^{n} w_j y_k^{(i)} \frac{\partial^2 H}{\partial x_j \partial x_k}(\mathbf{x}) \pmod{p^2}\\ &\equiv \mathbf{w}.\nabla H(\mathbf{x}) + p\sum_{j,k=1}^{n} \rho x_j y_k^{(i)} \frac{\partial^2 H}{\partial x_j \partial x_k}(\mathbf{x}) \pmod{p^2}\\ &\equiv \mathbf{w}.\nabla H(\mathbf{x}) + p(d-1)\rho\mathbf{y}^{(i)}.\nabla H(\mathbf{x}) \pmod{p^2}\\ &\equiv \mathbf{w}.\nabla H(\mathbf{x}) - (d-1)\rho H(\mathbf{x}) \pmod{p^2}\end{aligned}$$

for $i = 1, 2$. This therefore establishes the claim.

It is trivial to see that M is a lattice, since it is clearly closed under addition. Moreover if $\mathbf{y} \in \mathbb{Z}^3$, then $p^2\mathbf{y} \in \mathsf{M}$, by taking $\rho = 0$, whence M must have dimension 3. To compute $\det \mathsf{M}$ we observe that if we put $\mathbf{w} = \rho\mathbf{x}^{(1)} + p\mathbf{z}$, then we have

$$\begin{aligned}\mathbf{w}.\nabla H(\mathbf{x}^{(1)}) = \rho\mathbf{x}^{(1)}.\nabla H(\mathbf{x}^{(1)}) + p\mathbf{z}.\nabla H(\mathbf{x}^{(1)}) &= \rho d H(\mathbf{x}^{(1)}) + p\mathbf{z}.\nabla H(\mathbf{x}^{(1)})\\ &\equiv p\mathbf{z}.\nabla H(\mathbf{x}^{(1)}) \pmod{p^2},\end{aligned}$$

whence

$$\mathsf{M} = \{\mathbf{w} \in \mathbb{Z}^3 : \ \mathbf{w} = \rho\mathbf{x}^{(1)} + p\mathbf{z} \text{ some } \rho \in \mathbb{Z},\ \mathbf{z} \in \mathbb{Z}^3,\ p \mid \mathbf{z}.\nabla H(\mathbf{x}^{(1)})\}.$$

We now observe that we have an inclusion of lattices $p^2\mathbb{Z}^3 \subseteq \mathsf{M} \subseteq \mathbb{Z}^3$. Hence in order to calculate the determinant $\det \mathsf{M}$, which is equal to the index of M in $\mathbb{Z}^3$ as an additive subgroup, it will suffice to calculate the index $[\mathsf{M} : p^2\mathbb{Z}^3]$. Indeed we then have

$$\det \mathsf{M} = [\mathbb{Z}^3 : \mathsf{M}] = \frac{[\mathbb{Z}^3 : p^2\mathbb{Z}^3]}{[\mathsf{M} : p^2\mathbb{Z}^3]} = \frac{p^6}{[\mathsf{M} : p^2\mathbb{Z}^3]}. \tag{4.13}$$

We begin by considering the cosets of M modulo $p^2\mathbb{Z}^3$. In view of the coprimality constraint (4.12) there are p^2 possible values for $\mathbf{z}$ modulo p satisfying $p \mid \mathbf{z}.\nabla H(\mathbf{x}^{(1)})$. Moreover there are p^2 possible values for ρ modulo p^2. Finally, each value of $\mathbf{w}$ may be decomposed as $\mathbf{w} \equiv \rho\mathbf{x}^{(1)} + p\mathbf{z} \pmod{p^2}$ in p ways. It follows that M has exactly p^3 cosets modulo $p^2\mathbb{Z}^3$, and so (4.13) implies that $\det \mathsf{M} = p^6 p^{-3} = p^3$, as required.

For the final part of the lemma we note that if $\mathbf{t} \equiv \lambda\mathbf{x} \pmod{p}$, then $\mathbf{t} \equiv \lambda\mathbf{x}^{(1)} \pmod{p}$. Thus there is a vector $\mathbf{z} \in \mathbb{Z}^3$ such that $\mathbf{t} = \lambda\mathbf{x}^{(1)} + p\mathbf{z}$. It follows that

$$0 = H(\mathbf{t}) \equiv \lambda^d H(\mathbf{x}^{(1)}) + p\lambda^{d-1}\mathbf{z}.\nabla H(\mathbf{x}^{(1)}) \equiv p\lambda^{d-1}\mathbf{z}.\nabla H(\mathbf{x}^{(1)}) \pmod{p^2}.$$

Moreover we must have $\gcd(\lambda, p) = 1$ since $\mathbf{t}$ is primitive and $\mathbf{t} \equiv \lambda\mathbf{x} \pmod{p}$. This allows us to conclude that $p \mid \mathbf{z}.\nabla H(\mathbf{x}^{(1)})$, whence

$$\begin{aligned}\mathbf{t}.\nabla H(\mathbf{x}^{(1)}) &= \lambda\mathbf{x}^{(1)}.\nabla H(\mathbf{x}^{(1)}) + p\mathbf{z}.\nabla H(\mathbf{x}^{(1)})\\ &= \lambda d H(\mathbf{x}^{(1)}) + p\mathbf{z}.\nabla H(\mathbf{x}^{(1)}),\end{aligned}$$

which is congruent to 0 modulo p^2. Thus $\mathbf{t} \in \mathsf{M}$, as required. □

We can now complete the proof of Theorem 4.8. By our work above we may assume there is a prime $p \nmid \Delta$ in the range $cV^{\frac{1}{3}} \leqslant p \ll_d V^{\frac{1}{3}}$. The projective variety $F(\mathbf{x}) = 0$ has exactly $p + O(p^{\frac{1}{2}})$ points over $\mathbb{F}_p$, by Lemma 3.7. We aim to show that there are at most d points counted by $N(F;\mathbf{B})$ lying above each one. This will clearly suffice for the theorem. We therefore fix a vector $\mathbf{x} \in \mathbb{Z}^3_{\mathrm{prim}}$ with $F(\mathbf{x}) \equiv 0 \pmod{p}$, and note that (4.12) holds for $\mathbf{x}$ since $p \nmid \Delta$. Next we count vectors $\mathbf{w} \in \mathbb{Z}^3_{\mathrm{prim}}$ satisfying $F(\mathbf{w}) = 0$ and

$$|w_1| \leqslant B_1, \quad |w_2| \leqslant B_2, \quad |w_3| \leqslant B_3, \tag{4.14}$$

with $\mathbf{w} \equiv \rho\mathbf{x} \pmod{p}$, for some $\rho \in \mathbb{Z}$. According to Lemma 4.7 we will have $\mathbf{w} \in \mathsf{M}$.

We now consider the map

$$\phi(y_1, y_2, y_3) = (B_2B_3y_1\,,\ B_1B_3y_2\,,\ B_1B_2y_3).$$

Then $\phi(\mathsf{M})$ has determinant $V^2 \det \mathsf{M} = V^2p^3$. Let $s_1 \leqslant s_2 \leqslant s_3$ be the successive minima of $\phi(\mathsf{M})$, and let $\mathbf{b}^{(1)}, \mathbf{b}^{(2)}, \mathbf{b}^{(3)}$ be the corresponding minimal basis described in Lemma 4.3. It then follows that $s_3 \geqslant V^{\frac{2}{3}}p \gg_d V$. On the other hand, if (4.14) holds, then $\phi(\mathbf{w})$ has Euclidean length $\ll_d V$. Hence Lemma 4.3 implies that, if we choose the absolute constant c in (4.11) to be sufficiently large, the vector $\phi(\mathbf{w})$ must lie in the 2-dimensional lattice spanned by $\mathbf{b}^{(1)}$ and $\mathbf{b}^{(2)}$. Thus, given $\mathbf{x}$, all corresponding points $\mathbf{w}$ must lie not only on the curve $F = 0$ but also on a certain line. There are therefore at most d such points $\mathbf{w} \in \mathbb{Z}^3_{\mathrm{prim}}$ by Bézout's theorem, which thereby completes the proof of Theorem 4.8.

4.4 Diagonal plane curves

In this section we provide some estimates for the counting function associated to diagonal plane curves $C \subset \mathbb{P}^2$, given by equations of the form

$$a_1x_1^d + a_2x_2^d + a_3x_3^d = 0, \tag{4.15}$$

for $\mathbf{a} \in \mathbb{Z}^3$ and $d \in \mathbb{N}$. Certainly the estimates of the preceding sections are equipped to provide good upper bounds for $N_C(B)$. However, in the setting of diagonal curves, we seek versions of these estimates which actually get sharper as the size of the coefficients increases. These will prove useful in Chapter 6.

For the remainder of this section it will be convenient to write Z^n_* for the set of vectors $\mathbf{v} \in \mathbb{Z}^n_{\mathrm{prim}}$ such that $v_1 \cdots v_n \neq 0$. We clearly have $N_C(B) = \frac{1}{2}M_d(\mathbf{a}; B, B, B)$, where

$$M_d(\mathbf{a};\mathbf{B}) := \#\{\mathbf{x} \in \mathbb{Z}^3_{\mathrm{prim}} : a_1x_1^d + a_2x_2^d + a_3x_3^d = 0,\ |x_i| \leqslant B_i\}, \tag{4.16}$$

for non-zero integers a_1, a_2, a_3 and $\mathbf{B} = (B_1, B_2, B_3) \in \mathbb{R}^3_{>0}$. Let us begin with the situation for projective lines. The following result is an immediate consequence of Lemma 4.6.

Lemma 4.8. *Let $\mathbf{a} \in Z^3_*$. Then we have*

$$M_1(\mathbf{a}; \mathbf{B}) \ll 1 + \frac{B_1 B_2 B_3}{\max_i |a_i| B_i}.$$

Our next result features in joint work of Browning and Dietmann [29, § 4]. It establishes the existence of a small number of lattices, each of reasonably large determinant, that can be used to cover the integer solutions to the equation (4.15).

Lemma 4.9. *Let $d \geqslant 2$ and $\mathbf{a} \in \mathbb{Z}^3$ such that $a := |a_1 a_2 a_3| \neq 0$ and $\gcd(a_i, a_j) = 1$ for $1 \leqslant i < j \leqslant 3$. Let $\mathbf{x} \in \mathbb{Z}^3$ be a solution of (4.15). Then there exist lattices $\mathsf{\Lambda}_1, \ldots, \mathsf{\Lambda}_J \subseteq \mathbb{Z}^3$ such that*

(i) $J \leqslant 2d^{\omega(a)}$;

(ii) $\dim \mathsf{\Lambda}_j = 3$ *and* $\det \mathsf{\Lambda}_j \gg_d a^{\frac{2}{d}}$, *for each* $j \leqslant J$; *and*

(iii) $\mathbf{x} \in \bigcup_{j \leqslant J} \mathsf{\Lambda}_j$.

Proof. Let $p \mid a$, and write $\alpha_i = v_p(a_i)$. Since the integers a_1, a_2, a_3 are pairwise coprime, we may assume without loss of generality that $\alpha_1 = \alpha_2 = 0$ and $\alpha_3 \geqslant 1$. Let $\delta = v_p(d)$ be the p-adic order of d, and write

$$\gamma := \begin{cases} \delta + 1, & \text{if } p > 2 \text{ or } \delta = 0, \\ \delta + 2, & \text{if } p = 2 \text{ and } \delta \geqslant 1. \end{cases}$$

Let $\mathbf{x} \in \mathbb{Z}^3$ be such that (4.15) holds. We claim that there exist sublattices $\mathsf{M}_1, \ldots, \mathsf{M}_K \subseteq \mathbb{Z}^3$ with $K \leqslant 2^{\gamma-\delta-1} d$, such that

$$\dim \mathsf{M}_k = 3, \quad \det \mathsf{M}_k \geqslant p^{\frac{2\alpha_3}{d} - \gamma + 1},$$

for each $k \leqslant K$, and $\mathbf{x} \in \bigcup_{k \leqslant K} \mathsf{M}_k$. The Chinese remainder theorem will then produce at most $2d^{\omega(a)}$ integer sublattices overall, each of dimension 3 and determinant

$$\geqslant \prod_{p \mid a_1 a_2 a_3} p^{\frac{2 v_p(a)}{d} - \gamma + 1} = a^{\frac{2}{d}} \prod_{p \mid a_1 a_2 a_3} p^{-\gamma+1} \gg_d a^{\frac{2}{d}}.$$

This completes the proof of the lemma subject to the construction of the lattices $\mathsf{M}_1, \ldots, \mathsf{M}_K$.

Turning to the claim, let $\mathbf{x} \in \mathbb{Z}^3$ be such that (4.15) holds. Let us write $x_i = p^{\xi_i} x_i'$, for $i = 1, 2$, with $p \nmid x_1' x_2'$ and $\xi_1 \leqslant \xi_2$, say. Then we deduce that

$$a_1 p^{d\xi_1} {x_1'}^d + a_2 p^{d\xi_2} {x_2'}^d \equiv 0 \pmod{p^{\alpha_3}}.$$

There are now 3 possibilities to consider: either $\alpha_3 \leqslant d\xi_1$, or $d\xi_1 < \alpha_3 \leqslant d\xi_2$, or $d\xi_2 < \alpha_3$. The second case is plainly impossible. In the first case we may conclude that $\mathbf{x}$ belongs to the set of $\mathbf{x} \in \mathbb{Z}^3$ such that $p^{\lceil \frac{\alpha_3}{d} \rceil}$ divides x_1 and x_2. This defines an integer lattice of dimension 3 and determinant $\geqslant p^{\frac{2\alpha_3}{d}}$. Thus we may take $K = 1$ in this case.

Finally, in the third case, we must have $\xi_1 = \xi_2 = \xi$, say. But then it follows that

$$a_1 {x'_1}^d + a_2 {x'_2}^d \equiv 0 \pmod{p^{\alpha_3 - d\xi}}.$$

Suppose first that $\alpha_3 - d\xi < \gamma$. Then we have

$$p^{\frac{2\alpha_3}{d}} \leqslant p^{2\xi} p^{\frac{2(\gamma-1)}{d}} \leqslant p^{2\xi+\gamma-1}.$$

Since $\mathbf{x}$ lies on the lattice of determinant $p^{2\xi}$ that is determined by the conditions $p^\xi \mid x_1$ and $p^\xi \mid x_2$, we may clearly take $K = 1$ in this case also. Suppose now that $\alpha_3 - d\xi \geqslant \gamma$. Then we have

$$x^d + a_2 \overline{a_1} \equiv 0 \pmod{p^{\alpha_3 - d\xi}},$$

with $x = x'_1 \overline{x'_2}$, and where $\overline{b}$ denotes the multiplicative inverse of b modulo $p^{\alpha_3 - d\xi}$. We now appeal to Exercise 4.4. Thus for any $b \in \mathbb{Z}$ coprime to p, and any $k \geqslant \gamma$, the number of solutions to the congruence

$$x^d \equiv b \pmod{p^k}$$

is either 0 or $p^{\gamma-\delta-1} \gcd\big(d, p^\delta(p-1)\big)$. This therefore ensures the existence of $K \leqslant 2^{\gamma-\delta-1} d$ integers $\lambda_1, \ldots, \lambda_K$ such that

$$a_1 \lambda_k^d + a_2 \equiv 0 \pmod{p^{\alpha_3 - d\xi}},$$

for $1 \leqslant k \leqslant K$. In particular, the point $\mathbf{x} \in \mathbb{Z}^3$ in which we are interested must satisfy $x_1 = p^\xi x'_1$, $x_2 = p^\xi x'_2$ and

$$x'_1 \equiv \lambda_k x'_2 \pmod{p^{\alpha_3 - d\xi}},$$

for some $1 \leqslant k \leqslant K$. Assuming that $d \geqslant 2$, these conditions define a union of K lattices, each of dimension 3 and determinant $p^{\alpha_3 + 2\xi - d\xi} \geqslant p^{\frac{2\alpha_3}{d}}$. This completes the proof of the claim. □

We now have everything in place to establish the following analogue of Lemma 4.8 for diagonal forms of higher degree.

Lemma 4.10. *Let $d \geqslant 2$ and $\mathbf{a} \in Z_*^3$ such that $\gcd(a_i, a_j) = 1$. Then we have*

$$M_d(\mathbf{a}; \mathbf{B}) \ll_d \Big(1 + \frac{B_1 B_2 B_3}{|a_1 a_2 a_3|^{\frac{2}{d}}}\Big)^{\frac{1}{3}} d^{\omega(a_1 a_2 a_3)}.$$

Proof. Recall the definition (4.16) of the counting function $M_d(\mathbf{a};\mathbf{B})$ and let $a = |a_1a_2a_3|$. In view of Lemma 4.9, the points that we are interested in belong to a union of $J \leqslant 2d^{\omega(a)}$ lattices $\Lambda_1, \dots, \Lambda_J \subseteq \mathbb{Z}^3$, each of dimension 3 and determinant $\gg_d a^{\frac{2}{d}}$. Let us consider the overall contribution from the vectors belonging to one such lattice Λ_j, say.

Our argument now runs as in the proof of Lemma 4.6. Thus we note that the vectors $\mathbf{x}$ lying in the region $|x_i| \leqslant B_i$ also lie in the ellipsoid $\Sigma(\mathbf{x}) \leqslant 1$, with underlying matrix $3^{-1}\operatorname{Diag}(B_1^{-2}, B_2^{-2}, B_3^{-2})$ and volume $\ll B_1B_2B_3$. We carry out a unimodular transformation $\mathbf{T}$ of $\mathbb{R}^3$ taking the ellipsoid to a sphere of volume $V \ll B_1B_2B_3$, and Λ_j to a lattice $\mathbf{T}\Lambda_j$ of determinant $\gg_d a^{\frac{2}{d}}$. We will work with a minimal basis $\mathbf{b}^{(1)}, \mathbf{b}^{(2)}, \mathbf{b}^{(3)}$ for $\mathbf{T}\Lambda_j$, whose properties are recorded in Lemma 4.3. On making the change of variables

$$\mathbf{x} = \lambda_1\mathbf{b}^{(1)} + \lambda_2\mathbf{b}^{(2)} + \lambda_3\mathbf{b}^{(3)},$$

The equation (4.15) becomes $G_j(\lambda_1, \lambda_2, \lambda_3) = 0$, with G_j a non-singular ternary form of degree d, that is defined over $\mathbb{Z}$. We are now interested in counting integer solutions to this equation with $\lambda_i \ll V^{\frac{1}{3}}|\mathbf{b}^{(i)}|^{-1}$. The total number of such vectors is clearly

$$\ll_d 1 + \Big(\frac{V}{|\mathbf{b}^{(1)}||\mathbf{b}^{(2)}||\mathbf{b}^{(3)}|}\Big)^{\frac{1}{3}} \ll_d \Big(1 + \frac{B_1B_2B_3}{a^{\frac{2}{d}}}\Big)^{\frac{1}{3}},$$

by Theorem 4.8. On summing over the $J \leqslant 2d^{\omega(a)}$ lattices, this therefore completes the proof of the lemma. □

Exercises for Chapter 4

Exercise 4.1. Recall the definition (4.3) of $V_d \subset \mathbb{P}^3$. By considering the parameterisation

$$\mathbf{x} = (ut^{d-1}, vt^{d-1}, t^d, u^d + v^d),$$

for $u, v, t \in \mathbb{N}$, show that $N_{V_d}(B) \gg_d B^{\frac{3}{d}}$.

Exercise 4.2. Suppose that $\Lambda \subseteq \Gamma \subset \mathbb{Z}^n$ are two lattices of the same dimension. Show that $\det \Lambda \geqslant \det \Gamma$.

Exercise 4.3. If Λ and the basis $\mathbf{b}^{(1)}, \dots, \mathbf{b}^{(r)}$ are as in Lemma 4.3, then, for any integer $1 \leqslant k \leqslant r$, show that the determinant of the lattice

$$\mathsf{M} = \{\mathbf{x} \in \mathbb{Z}^n : \mathbf{x}.\mathbf{b}^{(1)} = \cdots = \mathbf{x}.\mathbf{b}^{(k)} = 0\}$$

has order of magnitude $\|\mathbf{b}^{(1)}\| \cdots \|\mathbf{b}^{(k)}\|$.

Exercise 4.4. Let $b \in \mathbb{Z}$ with $p \nmid b$. Let $k \geqslant \gamma$. Show that the number of solutions to the congruence

$$x^d \equiv b \pmod{p^k}$$

is either 0 or $p^{\gamma-\delta-1}\gcd\big(d, p^\delta(p-1)\big)$.

Chapter 5

$\mathbf{A}_1$ del Pezzo surface of degree 6

Let $S \subset \mathbb{P}^6$ be the surface cut out by the intersection of 9 quadrics

$$\begin{aligned} x_1^2 - x_2x_4 &= x_1x_5 - x_3x_4 = x_1x_3 - x_2x_5 = x_1x_6 - x_3x_5 \\ &= x_2x_6 - x_3^2 = x_4x_6 - x_5^2 = x_1^2 - x_1x_4 + x_5x_7 \\ &= x_1^2 - x_1x_2 - x_3x_7 = x_1x_3 - x_1x_5 + x_6x_7 = 0. \end{aligned} \tag{5.1}$$

This is the $\mathbf{A}_1$ del Pezzo surface of degree 6 that was introduced in (2.21). Any line in $\mathbb{P}^6$ is defined by the intersection of 5 hyperplanes. It is not hard to see that the equations

$$\begin{cases} x_1 = x_2 = x_3 = x_5 = x_6 = 0, & x_1 = x_3 = x_4 = x_5 = x_6 = 0, \\ x_3 = x_5 = x_6 = x_1 - x_4 = x_1 - x_2 = 0, \end{cases} \tag{5.2}$$

all define lines contained in S. Table 2.6 ensures that these are the only lines contained in S. We set U to be the open subset of S obtained by deleting the lines.

Our task in this chapter is to establish the existence of constants $c_1, c_2 \geqslant 0$ such that

$$N_U(B) = c_1 B(\log B)^3 + c_2 B(\log B)^2 + O\big(B \log B\big), \tag{5.3}$$

with

$$c_1 = \frac{\sigma_\infty}{144} \prod_p \Big(1 - \frac{1}{p}\Big)^4 \Big(1 + \frac{4}{p} + \frac{1}{p^2}\Big)$$

and

$$\sigma_\infty = 6 \int_{\{u,t,v \in \mathbb{R}:\ 0<u, ut^2, uv^2, |tv(t-v)| \leqslant 1\}} dt\, du\, dv.$$

This will therefore establish Theorem 2.4.

We begin by translating the counting problem into one in which none of the points counted by $N_U(B)$ are permitted to have zero coordinates.

Lemma 5.1. *We have*

$$N_U(B) = 2M(B) + O(B),$$

where $M(B)$ denotes the number of vectors $\mathbf{x} \in \mathbb{Z}^7_{\text{prim}}$ such that (5.1) *holds, with* $0 < |x_1|, x_2, x_3, x_4, |x_5|, x_6 \leqslant B$ *and* $|x_7| \leqslant B$.

Proof. In view of the fact that $\mathbf{x}$ and $-\mathbf{x}$ represent the same point in $\mathbb{P}^6$, we have

$$N_U(B) = \frac{1}{2}\#\{\mathbf{x} \in \mathbb{Z}^7_{\text{prim}} : |\mathbf{x}| \leqslant B, \text{ (5.1) holds but (5.2) does not}\}.$$

We need to consider the contribution to the right-hand side from points such that $x_i = 0$, for some $1 \leqslant i \leqslant 7$. Let us begin by considering the contribution from vectors $\mathbf{x} \in \mathbb{Z}^7_{\text{prim}}$ for which $x_1 = 0$. But then the equations in (5.1) imply that $x_2 x_4 = 0$. If $x_2 = 0$, it is straightforward to check that either $\mathbf{x}$ satisfies the first system of equations in (5.2), or else

$$x_1 = x_2 = x_3 = x_7 = 0, \quad x_4 x_6 = x_5^2.$$

Such points are therefore confined to a plane conic. We therefore obtain $O(B)$ points overall with $x_1 = x_2 = 0$ by Theorem 4.8. If on the other hand, $x_1 = x_4 = 0$, then a similar analysis shows that there are $O(B)$ points in this case too. In view of the first equation in (5.1), the contribution from vectors $\mathbf{x}$ such that $x_2 x_4 = 0$ is also $O(B)$.

Let us now consider the contribution from vectors $\mathbf{x}$ such that $x_3 = 0$ and $x_1 x_2 x_4 \neq 0$. It is easily checked that the only such vectors have $x_5 = x_6 = 0$ and $x_1 - x_4 = x_1 - x_2 = 0$, and so must lie on a line contained in S. Finally, arguing in a similar fashion, we see that there are no points contained in S with $x_5 x_6 = 0$ and $x_1 x_2 x_3 x_4 \neq 0$. We have therefore shown that

$$N_U(B) = \frac{1}{2}\#\{\mathbf{x} \in \mathbb{Z}^7_{\text{prim}} : x_1 \cdots x_6 \neq 0, \ |\mathbf{x}| \leqslant B, \text{ (5.1) holds}\} + O(B).$$

We would now like to restrict our attention to positive values of x_2, x_3, x_4, x_6. The equations for S imply that x_2, x_4, x_6 all share the same sign. On absorbing the minus sign into x_1 there is a clear bijection between solutions to (5.1) with $x_2, x_4, x_6 < 0$ and solutions with $x_2, x_4, x_6 > 0$. We choose to count the latter. Arguing similarly, by absorbing the minus signs into x_5 and x_7, we see that there is a bijection between the solutions to (5.1) with $x_3 < 0$ and $x_2, x_4, x_6 > 0$, and the solutions with $x_2, x_3, x_4, x_6 > 0$. Fixing our attention on the latter set of points, we therefore complete the proof of Lemma 5.1. □

5.1 Passage to the universal torsor

Let $\widetilde{S}$ denote the minimal desingularisation of the surface S. By determining the Cox ring, Derenthal [53] has calculated the universal torsor above $\widetilde{S}$. In this setting it is defined by a single equation

$$s_1 y_1 - s_2 y_2 + s_3 y_3 = 0, \tag{5.4}$$

embedded in $\mathbb{A}^7$. In particular one of the variables does not appear explicitly in the equation. We proceed to show how $N_U(B)$ can be related to a counting problem for integer points on (5.4). Our deduction of this fact is completely elementary, and is based on an analysis of the integer solutions to the system of equations (5.1). It is still somewhat mysterious as to how or why this rather low-brow process should ultimately lead to the same outcome!

Given $s_0 \in \mathbb{R}$ and $\mathbf{s}, \mathbf{y} \in \mathbb{R}^3$, define

$$\Psi(s_0, \mathbf{s}, \mathbf{y}) := \max\left\{|s_0^3 s_1^2 s_2^2 s_3^2|,\ |y_1 y_2 y_3|,\ |s_0 s_1^2 y_1^2|,\ |s_0 s_2^2 y_2^2|\right\}. \tag{5.5}$$

We are now ready to record our translation of the problem to the universal torsor.

Lemma 5.2. *We have*

$$N_U(B) = 2\#\left\{(s_0, \mathbf{s}, \mathbf{y}) \in \mathbb{Z}^7 : \begin{array}{l} \Psi(s_0, \mathbf{s}, \mathbf{y}) \leqslant B,\ (5.4)\ \textit{holds}, \\ s_0, s_1, s_2, s_3, y_1, |y_2| > 0, \\ \gcd(y_i, s_0 s_j s_k) = 1, \\ \gcd(s_i, s_j) = 1 \end{array}\right\} + O(B),$$

with i, j, k *a permutation of* $1, 2, 3$ *in the coprimality conditions.*

Proof. Let $\mathbf{x} \in \mathbb{Z}_{\mathrm{prim}}^7$ be a vector counted by $M(B)$, as defined in the statement of Lemma 5.1. The general solution of the equation $xy = z^2$ is

$$x = a^2 c, \quad y = b^2 c, \quad z = abc,$$

with $|\mu(c)| = 1$. Applying this in the first equation in (5.1), we conclude that

$$x_1 = a_1 a_2 a_4, \quad x_2 = a_2^2 a_1, \quad x_4 = a_4^2 a_1,$$

for integers a_1, a_2, a_4 such that $a_1, a_2 > 0$ and

$$|\mu(a_1)| = \gcd(a_1 \gcd(a_2, a_4)^2, x_3, x_5, x_6, x_7) = 1.$$

Inserting this into the equation $x_2 x_6 = x_3^2$ we deduce that $a_1 a_2 \mid x_3$, whence

$$x_3 = a_1 a_2 a_3, \quad x_6 = a_1 a_3^2,$$

for a positive integer a_3 such that

$$|\mu(a_1)| = \gcd\Big(a_1 \gcd(a_2, a_4)^2, a_1 a_3 \gcd(a_2, a_3), x_5, x_7\Big) = 1.$$

Substituting this into the equation $x_4 x_6 = x_5^2$, we deduce that

$$x_5 = a_1 a_3 a_4,$$

with

$$|\mu(a_1)| = \gcd(a_1, x_7) = \gcd(a_2, a_3, a_4, x_7) = 1. \tag{5.6}$$

Note that the second equation in (5.1) implies that x_1, x_5 must share the same sign, which here is the sign of a_4.

The equations $x_1x_5 = x_3x_4$, $x_1x_3 = x_2x_5$ and $x_1x_6 = x_3x_5$ reveal no new information. Turning instead to the equation $x_1^2 = x_1x_4 - x_5x_7$, we obtain

$$a_1a_2^2a_4 = a_1a_2a_4^2 - a_3x_7. \tag{5.7}$$

The coprimality conditions imply that $a_1 \mid a_3$. Moreover, we deduce from this equation that $a_2a_4 \mid a_1^{-1}a_3x_7$. We may therefore write

$$a_2 = a_{23}a_{27}, \quad a_4 = a_{43}a_{47},$$

for integers a_{2i}, a_{4i}, with $i = 3, 7$, such that $a_{2i}, a_{43}, |a_{47}| > 0$, and

$$a_1a_{23}a_{43} \mid a_3, \quad a_{27}a_{47} \mid x_7.$$

Thus there exist further integers b_3, a_7 with $b_3 > 0$, such that

$$a_3 = a_1a_{23}a_{43}b_3, \quad x_7 = a_{27}a_{47}a_7,$$

with (5.6) and (5.7) becoming

$$|\mu(a_1)| = \gcd(a_1, a_{27}a_{47}a_7) = \gcd(a_{23}a_{27}, a_{23}a_{43}b_3, a_{43}a_{47}, a_{27}a_{47}a_7) = 1,$$

and

$$a_{23}a_{27} = a_{43}a_{47} - b_3a_7,$$

respectively. The final two equations are redundant.

Let us write d for the highest common factor of a_{23}, a_{43}, b_3. Thus

$$a_{23} = da'_{23}, \quad a_{43} = da'_{43}, \quad b_3 = db'_3,$$

for positive integers $d, a'_{23}, a'_{43}, b'_3$. On making these substitutions the equation remains the same, but with appropriate accents added, whereas the coprimality conditions become

$$\begin{aligned} |\mu(a_1)| = \gcd(da_1, a_{27}a_{47}a_7) &= \gcd(a'_{23}a_{27}, a'_{23}a'_{43}b'_3, a'_{43}a_{47}, a_{27}a_{47}a_7) \\ &= \gcd(a'_{23}, a'_{43}, b'_3) \\ &= 1. \end{aligned}$$

Now any $n \in \mathbb{N}$ can be written uniquely in the form $n = ab^2$ for $a, b \in \mathbb{N}$ such that $|\mu(a)| = 1$. We may therefore make the change of variables

$$(s_0; s_1, s_2, s_3; y_1, y_2, y_3) = (a_1d^2; a'_{23}, a'_{43}, b'_3; a_{27}, a_{47}, a_7).$$

Bringing everything together, we have therefore established the existence of $(s_0, \mathbf{s}, \mathbf{y}) \in \mathbb{Z}^7$ such that (5.4) holds, with

$$s_0, s_1, s_2, s_3, y_1, |y_2| > 0, \tag{5.8}$$

and

$$\gcd(s_0, y_1y_2y_3) = \gcd(s_1, s_2, s_3) = \gcd(s_1y_1, s_2y_2, s_1s_2s_3, y_1y_2y_3) = 1.$$

Once combined with (5.4), it is easy to check that the latter coprimality conditions are equivalent to the conditions

$$\begin{cases} \gcd(y_1, s_0s_2s_3) = \gcd(y_2, s_0s_1s_3) = \gcd(y_3, s_0s_1s_2) = 1, \\ \gcd(s_1, s_2) = \gcd(s_1, s_3) = \gcd(s_2, s_3) = 1, \end{cases} \tag{5.9}$$

that appear in the statement of the lemma.

At this point we may summarise our argument as follows. Let $\mathscr{T} \subset \mathbb{Z}^7$ denote the set of $(s_0, \mathbf{s}, \mathbf{y}) \in \mathbb{Z}^7$ such that (5.4), (5.8) and (5.9) hold. Then for any primitive vector $\mathbf{x}$ counted by $M(B)$, we have shown that there exists $(s_0, \mathbf{s}, \mathbf{y}) \in \mathscr{T}$ such that

$$\begin{cases} x_1 = s_0s_1s_2y_1y_2, \\ x_2 = s_0s_1^2y_1^2, \\ x_3 = s_0^2s_1^2s_2s_3y_1, \\ x_4 = s_0s_2^2y_2^2, \\ x_5 = s_0^2s_1s_2^2s_3y_2, \\ x_6 = s_0^3s_1^2s_2^2s_3^2, \\ x_7 = y_1y_2y_3. \end{cases}$$

Conversely, we leave it as an exercise to check that any $(s_0, \mathbf{s}, \mathbf{y}) \in \mathscr{T}$ produces a primitive point $\mathbf{x} \in \mathbb{Z}^7$ such that (5.1) holds, with

$$|x_1|, x_2, x_3, x_4, |x_5|, x_6 > 0.$$

Note that $|s_0s_1s_2y_1y_2| = \sqrt{|s_0s_1^2y_1^2|}\sqrt{|s_0s_2^2y_2^2|}$, and furthermore,

$$|s_0^2s_1s_2s_3s_iy_i| = \sqrt{|s_0^3s_1^2s_2^2s_3^2|}\sqrt{|s_0s_i^2y_i^2|}.$$

We may now conclude that $M(B)$ is equal to the number of $(s_0, \mathbf{s}, \mathbf{y}) \in \mathscr{T}$ such that $\Psi(s_0, \mathbf{s}, \mathbf{y}) \leqslant B$, where Ψ is given by (5.5). Once inserted into Lemma 5.1, this completes the proof of Lemma 5.2. □

At first glance it might seem odd that the height restriction $|s_0s_3^2y_3^2| \leqslant B$ does not explicitly appear in the lemma. However, (5.4) implies that

$$0 < s_1y_1 = s_2y_2 - s_3y_3$$

for any $(s_0, \mathbf{s}, \mathbf{y}) \in \mathscr{T}$, whence the restriction $\Psi(s_0, \mathbf{s}, \mathbf{y}) \leqslant B$ is plainly equivalent to $\max\{|s_0s_3^2y_3^2|, \Psi(s_0, \mathbf{s}, \mathbf{y})\} \leqslant B$. We have preferred not to include it explicitly in the statement of Lemma 5.2 however.

5.2 The asymptotic formula

Our starting point is Lemma 5.2. Let $T(B)$ denote the cardinality on the right-hand side, so that $N_U(B) = 2T(B) + O(B)$. Once taken together with (5.4), the height condition $\Psi(s_0, \mathbf{s}, \mathbf{y}) \leqslant B$ is equivalent to

$$\max_{i=1,2} \{|s_0^3 s_1^2 s_2^2 s_3^2|, |s_0 s_i^2 y_i^2|, \ |y_1 y_2 (s_1 y_1 - s_2 y_2) s_3^{-1}|\} \leqslant B.$$

Define

$$X_0 := \left(\frac{s_0^3 s_1^2 s_2^2 s_3^2}{B}\right)^{\frac{1}{3}}, \quad X_i := \left(\frac{s_1 s_2 s_3 B}{s_i^3}\right)^{\frac{1}{3}},$$

for $i = 1, 2$. Then the height conditions above can be rewritten as

$$X_0 \leqslant 1, \quad |f_1(y_1)| \leqslant 1, \quad |f_2(y_2)| \leqslant 1, \quad |g(y_1, y_2)| \leqslant 1,$$

where

$$f_i(y) := X_0 \Big(\frac{y}{X_i}\Big)^2, \quad g(y_1, y_2) := \frac{y_1 y_2}{X_1 X_2}\Big(\frac{y_1}{X_1} - \frac{y_2}{X_2}\Big),$$

for $i = 1, 2$.

In order to count solutions to the equation (5.4), our plan will be to view the equation as a congruence

$$s_1 y_1 - s_2 y_2 \equiv 0 \pmod{s_3},$$

which has the effect of automatically taking care of the summation over y_3. In order to make this approach viable we will first need to extract the coprimality conditions on the y_3 variable.

Define the set

$$\mathscr{S} := \{(s_0, \mathbf{s}) \in \mathbb{N}^4 : \gcd(s_i, s_j) = 1, \ X_0 \leqslant 1\}, \tag{5.10}$$

with i, j generic indices from the set $\{1, 2, 3\}$. We now apply Möbius inversion to remove the coprimality condition $\gcd(y_3, s_0 s_1 s_2) = 1$. This gives

$$T(B) = \sum_{(s_0,\mathbf{s})\in\mathscr{S}} \sum_{k_3 | s_0 s_1 s_2} \mu(k_3) \# \left\{ \mathbf{y} \in \mathbb{Z}^3 : \begin{array}{l} \gcd(y_1, s_0 s_2 s_3) = 1, \\ \gcd(y_2, s_0 s_1 s_3) = 1, \\ y_1, |y_2| > 0, \\ s_1 y_1 - s_2 y_2 + k_3 s_3 y_3 = 0, \\ |f_i(y_i)| \leqslant 1, \ |g(y_1, y_2)| \leqslant 1 \end{array} \right\}.$$

Now it is clear that the summand vanishes unless $\gcd(k_3, s_1 s_2) = 1$. Hence

$$T(B) = \sum_{(s_0,\mathbf{s})\in\mathscr{S}} \sum_{\substack{k_3 | s_0 \\ \gcd(k_3, s_1 s_2) = 1}} \mu(k_3) S_{k_3}(B), \tag{5.11}$$

where

$$S_{k_3}(B) := \#\left\{ y_1, y_2 \in \mathbb{Z} : \begin{array}{l} \gcd(y_1, s_0 s_2 s_3) = 1, \\ \gcd(y_2, s_0 s_1 s_3) = 1, \\ y_1, |y_2| > 0, \\ s_1 y_1 \equiv s_2 y_2 \pmod{k_3 s_3}, \\ |f_i(y_i)| \leqslant 1, \ |g(y_1, y_2)| \leqslant 1 \end{array} \right\}.$$

Clearly $S_{k_3}(B)$ depends on the parameters s_0 and $\mathbf{s}$, in addition to k_3 and B.

To estimate $S_{k_3}(B)$ we will fix y_2 and apply Exercise 5.1 to handle the summation over y_1. Before this we must use Möbius inversion to remove the coprimality condition $\gcd(y_1, s_0 s_2 s_3) = 1$ from the summand. Thus we find that

$$S_{k_3}(B) = \sum_{k_1 \mid s_0 s_2 s_3} \mu(k_1) \# \left\{ y_1, y_2 \in \mathbb{Z} : \begin{array}{l} \gcd(y_2, s_0 s_1 s_3) = 1, \\ k_1 s_1 y_1 \equiv s_2 y_2 \pmod{k_3 s_3}, \\ |f_1(k_1 y_1)| \leqslant 1, \ |f_2(y_2)| \leqslant 1, \\ |g(k_1 y_1, y_2)| \leqslant 1, y_1, |y_2| > 0 \end{array} \right\}.$$

In view of the other coprimality conditions, the summand plainly vanishes unless $\gcd(k_1, k_3 s_3) = 1$. We may therefore write $\rho \in \mathbb{Z}$ for the (unique) inverse of $k_1 s_1$ modulo $k_3 s_3$, whence

$$S_{k_3}(B) = \sum_{\substack{k_1 \mid s_0 s_2 \\ \gcd(k_1, k_3 s_3) = 1}} \mu(k_1) S_{k_1, k_3}(B), \tag{5.12}$$

with

$$S_{k_1,k_3}(B) := \sum_{\substack{y_2 \in \mathbb{Z}_{\neq 0}: \ |f_2(y_2)| \leqslant 1 \\ \gcd(y_2, s_0 s_1 s_3) = 1}} \# \left\{ y_1 \in \mathbb{N} : \begin{array}{l} y_1 \equiv \rho s_2 y_2 \pmod{k_3 s_3}, \\ |f_1(k_1 y_1)| \leqslant 1, \\ |g(k_1 y_1, y_2)| \leqslant 1 \end{array} \right\}.$$

An application of Exercise 5.1 now reveals that

$$S_{k_1,k_3}(B) = \sum_{\substack{y_2 \in \mathbb{Z}_{\neq 0}: \ |f_2(y_2)| \leqslant 1 \\ \gcd(y_2, s_0 s_1 s_3) = 1}} \Big(\frac{X_1 F_1(X_0, \frac{y_2}{X_2})}{k_1 k_3 s_3} + O(1) \Big), \tag{5.13}$$

where

$$F_1(u, v) := \int_{\{t \in \mathbb{R}: \ t > 0, \ |ut^2|, |tv(t-v)| \leqslant 1\}} \mathrm{d}t.$$

We now need to show that once summed over all $s_0, \mathbf{s}, y_2$, the error term in (5.13) makes a satisfactory overall contribution to the error term in (5.3). Using the fact that $\sum_{k|n} |\mu(k)| = 2^{\omega(n)}$, we find that this contribution is

$$\begin{aligned} &\ll \sum_{(s_0,\mathbf{s})\in\mathscr{S}} \frac{4^{\omega(s_0)}2^{\omega(s_2)}X_2}{X_0^{\frac{1}{2}}} = B^{\frac{1}{2}} \sum_{(s_0,\mathbf{s})\in\mathscr{S}} \frac{4^{\omega(s_0)}2^{\omega(s_2)}}{s_0^{\frac{1}{2}}s_2} \\ &\ll B \sum_{\substack{s_0,s_1,s_2\in\mathbb{N}\\ s_0^3s_1^2s_2^2\leqslant B}} \frac{4^{\omega(s_0)}2^{\omega(s_2)}}{s_0^2s_1s_2^2} \\ &\ll B\log B. \end{aligned}$$

This is satisfactory and so we may henceforth ignore the error term in the above estimate for $S_{k_1,k_3}(B)$.

Define the arithmetic function

$$\phi^*(n) := \prod_{p|n}\Big(1-\frac{1}{p}\Big),$$

where by common convention the product is over distinct prime divisors of n. It will be useful to note that

$$\phi^*(mn) = \frac{\phi^*(m)\phi^*(n)}{\phi^*(\gcd(m,n))}, \tag{5.14}$$

for any $m,n\in\mathbb{N}$. We must now sum over the variable y_2.

Returning to (5.13) and applying Exercise 5.2 we deduce that

$$S_{k_1,k_3}(B) = \frac{\phi^*(s_0s_1s_3)X_1X_2F_2(X_0)}{k_1k_3s_3} + O\Big(\frac{2^{\omega(s_0s_1s_3)}X_1}{k_1k_3s_3X_0^{\frac{1}{2}}}\Big), \tag{5.15}$$

where

$$F_2(u) := \int_{\{t,v\in\mathbb{R}:\ t>0,\ |ut^2|,|uv^2|,|tv(t-v)|\leqslant 1\}} \mathrm{d}t\,\mathrm{d}v.$$

We must now estimate the overall contribution to $N_U(B)$ from the error term in this estimate, once summed up over the remaining variables. This gives

$$\begin{aligned} \ll \sum_{(s_0,\mathbf{s})\in\mathscr{S}} \frac{4^{\omega(s_0)}2^{\omega(s_2)}2^{\omega(s_0s_1s_3)}X_1}{s_3X_0^{\frac{1}{2}}} &\ll B^{\frac{1}{2}} \sum_{\substack{s_0,s_1,s_2,s_3\in\mathbb{N}\\ s_0^3s_1^2s_2^2s_3^2\leqslant B}} \frac{8^{\omega(s_0)}2^{\omega(s_1s_2s_3)}}{s_0^{\frac{1}{2}}s_1s_3} \\ &\ll B\log B, \end{aligned}$$

by summing over s_2 first. This is satisfactory for (5.3), and so we may henceforth ignore the error term in (5.15).

Bringing together Exercise 5.3 with (5.11), (5.12) and (5.15), we conclude that

$$T(B) = \sum_{(s_0,\mathbf{s})\in\mathscr{S}} \sum_{\substack{k_3|s_0\\ \gcd(k_3,s_1s_2)=1}} \frac{\mu(k_3)}{k_3}\cdot\frac{\phi^*(s_0s_2)\phi^*(s_0s_1s_3)}{\phi^*(\gcd(k_3s_3,s_0s_2))}\cdot\frac{X_1X_2F_2(X_0)}{s_3},$$

where $\mathscr{S}$ is given by (5.10). It is clear that $\gcd(k_3s_3, s_0s_2) = \gcd(k_3s_3, s_0)$. Let us define the arithmetic function

$$\vartheta(s_0, \mathbf{s}) = \frac{\phi^*(s_0s_2)\phi^*(s_0s_1s_3)}{\phi^*(\gcd(s_0, s_3))} \sum_{\substack{k_3|s_0 \\ \gcd(k_3,s_1s_2)=1}} \frac{\mu(k_3)}{k_3} \cdot \frac{\phi^*(\gcd(k_3, s_0, s_3))}{\phi^*(\gcd(k_3, s_0))}$$

when $\gcd(s_i, s_j) = 1$ for $1 \leqslant i < j \leqslant 3$, and $\vartheta(s_0, \mathbf{s}) = 0$ otherwise. It follows from (5.14) that

$$\begin{aligned}
\vartheta(s_0, \mathbf{s}) &= \frac{\phi^*(s_0s_2)\phi^*(s_0s_1s_3)}{\phi^*(\gcd(s_0, s_3))} \prod_{\substack{p|\gcd(s_0,s_3) \\ p\nmid s_1s_2}} \Big(1 - \frac{1}{p}\Big) \prod_{\substack{p|s_0 \\ p\nmid s_1s_2s_3}} \Big(\frac{1-\frac{2}{p}}{1-\frac{1}{p}}\Big) \\
&= \phi^*(s_0s_2)\phi^*(s_0s_1s_3) \prod_{\substack{p|s_0 \\ p\nmid s_1s_2s_3}} \Big(\frac{1-\frac{2}{p}}{1-\frac{1}{p}}\Big) \\
&= \phi^*(s_0)\phi^*(s_1s_2s_3) \prod_{\substack{p|s_0 \\ p\nmid s_1s_2s_3}} \Big(1 - \frac{2}{p}\Big),
\end{aligned}$$

when $\gcd(s_i, s_j) = 1$ for $1 \leqslant i < j \leqslant 3$. In particular we have

$$0 \leqslant \vartheta(s_0, \mathbf{s}) \leqslant 1. \tag{5.16}$$

On recalling the definitions of X_1, X_2, we conclude that

$$T(B) = B^{\frac{2}{3}} \sum_{n \leqslant B} \Delta(n) F_2\Big(\Big(\frac{n}{B}\Big)^{\frac{1}{3}}\Big), \tag{5.17}$$

where

$$\Delta(n) := \sum_{n=s_0^3s_1^2s_2^2s_3^2} \frac{\vartheta(s_0, \mathbf{s})}{(s_1s_2s_3)^{\frac{1}{3}}}.$$

5.3 Perron's formula

It remains to carry out the summation over n in (5.17). We shall do so by estimating $\sum_{n\leqslant B} \Delta(n)$ first, before combining it with partial summation to estimate (5.17). There is a powerful tool in analytic number theory for estimating general sums of the sort

$$S_h(x) := \sum_{n \leqslant x} h(n), \tag{5.18}$$

as $x \to \infty$, for multiplicative arithmetic functions $h : \mathbb{N} \to \mathbb{C}$. The key is to understand the analytic properties of the associated Dirichlet series

$$H(s) := \sum_{n=1}^{\infty} \frac{h(n)}{n^s}, \tag{5.19}$$

for $s \in \mathbb{C}$ with real part sufficiently large to ensure the absolute convergence of the series. We have already met this sort of phenomenon in our discussion of height zeta functions in Section 2.6.

Let

$$\delta(x) := \begin{cases} 0, & \text{if } 0 < x < 1, \\ \frac{1}{2}, & \text{if } x = 1, \\ 1, & \text{if } x > 1. \end{cases}$$

It can be shown that

$$\delta(x) = \frac{1}{2\pi i} \int_{c-i\infty}^{c+i\infty} \frac{x^s}{s} \mathrm{d}s,$$

for any $c, x > 0$. We would like a version of this result which deals with a truncated range of integration. This is provided by the following well-known result.

Lemma 5.3. *For any $c, x, T > 0$ we have*

$$\left| \frac{1}{2\pi i} \int_{c-iT}^{c+iT} \frac{x^s}{s} \mathrm{d}s - \delta(x) \right| < \begin{cases} x^c \min\{1, T^{-1}|\log x|^{-1}\}, & \textit{if } x \neq 1, \\ \frac{c}{T}, & \textit{if } x = 1. \end{cases}$$

Proof. Suppose first that $0 < x < 1$. Consider the rectangular contour $\mathscr{C}$ joining $c - iT, c + iT, d + iT, d - iT$ for $d > c$. By Cauchy's theorem we have

$$\frac{1}{2\pi i} \int_{\mathscr{C}} \frac{x^s}{s} \mathrm{d}s = 0 = \delta(x).$$

To prove the theorem we need to estimate the three integrals

$$\frac{1}{2\pi i} \int_{c\pm iT}^{d\pm iT} \frac{x^s}{s} \mathrm{d}s, \quad \frac{1}{2\pi i} \int_{d-iT}^{d+iT} \frac{x^s}{s} \mathrm{d}s,$$

which we denote by $I_\pm, J$, respectively. Now

$$|I_\pm| \leqslant \frac{1}{2\pi T} \int_c^d x^\sigma \mathrm{d}\sigma \leqslant \frac{x^c}{T|\log x|},$$

for $d \to \infty$. Next we note that

$$|J| \leqslant \frac{x^d T}{d},$$

which goes to zero as $d \to \infty$, since $0 < x < 1$.

This proves one of the two stated inequalities in the case $0 < x < 1$. To prove the other one, we consider the circle $(c^2 + T^2)^{\frac{1}{2}}$ centred on the origin. This circle passes through $c - iT$ and $c + iT$. We can therefore replace the vertical line integral under consideration by a circular path on the right side of the line segment joining $c - iT$ to $c + iT$. The integral is easily estimated as

$$\left| \frac{1}{2\pi i} \int_{c-iT}^{c+iT} \frac{x^s}{s} \mathrm{d}s \right| \leqslant \frac{1}{2\pi} \cdot \pi T \cdot \frac{x^c}{T} < x^c,$$

since $|x^s| \leqslant x^c$ on the circular path.

The proof when $x > 1$ is similar but uses a rectangle or a circular arc to the left. The details are left to the reader, who should note that the contour then encounters the pole at $s = 0$, with residue $1 = \delta(x)$.

Finally, for the case $x = 1$, we have

$$\frac{1}{2\pi i}\int_{c-iT}^{c+iT}\frac{ds}{s} = \frac{c}{\pi}\int_0^T \frac{dt}{c^2+t^2} = \frac{1}{\pi}\int_0^{\frac{T}{c}}\frac{du}{1+u^2} = \frac{1}{2} - \frac{1}{\pi}\int_{\frac{T}{c}}^{\infty}\frac{du}{1+u^2}.$$

The final integral is less than $\frac{c}{T}$, which therefore completes the proof of the lemma. □

We are now ready to record the truncated version of Perron's formula that we will need. Recall the definition (5.19) of the Dirichlet series $H(s)$ that is associated to a multiplicative arithmetic function $h(n)$ and the corresponding summatory function (5.18).

Lemma 5.4. *Suppose that $H(s)$ has abscissa of absolute convergence σ_a. Then for any $x, T > 0$ and $c > \sigma_a$, with $x \notin \mathbb{Z}$, we have*

$$S_h(x) = \frac{1}{2\pi i}\int_{c-iT}^{c+iT} H(s)\frac{x^s}{s}ds + O\left(\frac{x^c}{T} + \sum_{\frac{x}{2}\leqslant n\leqslant\frac{3x}{2}} |h(n)| \min\left\{1, \frac{1}{T|\log\frac{x}{n}|}\right\}\right).$$

Proof. It follows from a straightforward application of Lemma 5.3 that

$$S_h(x) = \frac{1}{2\pi i}\int_{c-iT}^{c+iT} H(s)\frac{x^s}{s}ds + O\left(\sum_{n=1}^{\infty}\left(\frac{x}{n}\right)^c |h(n)| \min\left\{1, \frac{1}{T|\log\frac{x}{n}|}\right\}\right).$$

The error term here is $O(x^cE)$, with

$$E = \sum_{n=1}^{\infty}\frac{|h(n)|}{n^c}\min\left\{1, \frac{1}{T|\log\frac{x}{n}|}\right\}.$$

Moreover, we have $\sum_{n=1}^{\infty}|h(n)|n^{-c} \ll 1$, by assumption. Let us break the summation over n into intervals $[1, \frac{x}{2}) \cup [\frac{x}{2}, \frac{3x}{2}] \cup (\frac{3x}{2}, \infty)$. We denote by E_i the overall contribution to E from the ith interval, for $1 \leqslant i \leqslant 3$.

Since $|\log\frac{x}{n}| \gg 1$ for $n \in [1, \frac{x}{2}) \cup (\frac{3x}{2}, \infty)$, so it follows that

$$E_i \ll \frac{1}{T}$$

for $i = 1, 3$. This is satisfactory for the lemma. Finally we note that

$$E_2 \ll x^{-c}\sum_{\frac{x}{2}\leqslant n\leqslant\frac{3x}{2}} |h(n)| \min\left\{1, \frac{1}{T|\log\frac{x}{n}|}\right\},$$

which is also satisfactory. This therefore completes the proof of the lemma. □

We will use Perron's formula, in the form of Lemma 5.4, to estimate the sum $\sum_{n \leqslant B} \Delta(n)$, before combining it with partial summation to estimate (5.17). Consider the Dirichlet series $D(s) := \sum_{n=1}^{\infty} \Delta(n) n^{-s}$. We have

$$D\Big(s+\frac{1}{3}\Big) = \sum_{s_0,s_1,s_2,s_3=1}^{\infty} \frac{\vartheta(s_0,\mathbf{s})}{s_0^{3s+1} s_1^{2s+1} s_2^{2s+1} s_3^{2s+1}}.$$

It is straightforward to check that $D(s+\frac{1}{3}) = \prod_p a_p(s)$, with

$$a_p(s) = 1 + \frac{3(1-\frac{1}{p})}{p^{2s+1}(1-\frac{1}{p^{2s+1}})} + \frac{(1-\frac{1}{p})(1-\frac{2}{p})}{p^{3s+1}(1-\frac{1}{p^{3s+1}})} + \frac{3(1-\frac{1}{p})^2}{p^{5s+2}(1-\frac{1}{p^{2s+1}})(1-\frac{1}{p^{3s+1}})}.$$

Hence $D(s+\frac{1}{3}) = E_1(s)E_2(s)$, where $E_1(s) = \zeta(2s+1)^3\zeta(3s+1)$ and

$$E_2(s) = \frac{D(s+\frac{1}{3})}{\zeta(2s+1)^3\zeta(3s+1)} = \prod_p \Big(1 + O\Big(\frac{1}{p^{\min\{6\sigma+2,2\sigma+2\}}}\Big)\Big) \tag{5.20}$$

on the half-plane $\Re e(s) > -\frac{1}{6}$. In particular, $E_1(s)$ has a meromorphic continuation to all of $\mathbb{C}$ with a pole of order 4 at $s=0$, and $E_2(s)$ is holomorphic and bounded on the half-plane $\Re e(s) > -\frac{1}{6}$.

The abscissa of absolute convergence for $D(s)$ is $\frac{1}{3}$. We may therefore take $c = \frac{1}{3}+\varepsilon$ for any $\varepsilon > 0$ in Lemma 5.4. It follows from (5.16) that the error term in Lemma 5.4 is

$$\ll \frac{B^{\frac{1}{3}+\varepsilon}}{T} + \sum_{s_1,s_2,s_3 \ll B^{\frac{1}{2}}} \frac{1}{(s_1s_2s_3)^{\frac{1}{3}}} \sum_{ay \leqslant s_0 \leqslant by} \min\Big\{1, \frac{1}{T|\log\frac{s_0}{y}|}\Big\},$$

where $y := (s_1s_2s_3)^{-\frac{2}{3}}B^{\frac{1}{3}}$ and a,b are absolute constants. Writing $N = [ay]$ and $s_0 = N+m$, for $0 \leqslant m \leqslant by$, we see that

$$\Big|\log\frac{s_0}{y}\Big| \gg \frac{m}{y}.$$

Thus the error term in Lemma 5.4 is

$$\begin{aligned}
&\ll \frac{B^{\frac{1}{3}+\varepsilon}}{T} + \sum_{s_1,s_2,s_3 \ll B^{\frac{1}{2}}} \frac{1}{(s_1s_2s_3)^{\frac{1}{3}}} \sum_{0 \leqslant m \leqslant by} \min\Big\{1, \frac{y}{mT}\Big\} \\
&\ll_\varepsilon \frac{B^{\frac{1}{3}+\varepsilon}}{T} + \frac{B^{\frac{1}{3}}\log B}{T} \sum_{s_1,s_2,s_3 \ll B^{\frac{1}{2}}} \frac{1}{s_1s_2s_3} \\
&\ll_\varepsilon \frac{B^{\frac{1}{3}+\varepsilon}}{T}.
\end{aligned}$$

We conclude that

$$\sum_{n\leqslant B}\Delta(n)=\frac{1}{2\pi i}\int_{c-iT}^{c+iT}E_1\Big(s-\frac{1}{3}\Big)E_2\Big(s-\frac{1}{3}\Big)\frac{B^s}{s}\mathrm{d}s+O_\varepsilon\Big(\frac{B^{\frac{1}{3}+\varepsilon}}{T}\Big),$$

provided that $B\notin\mathbb{Z}$ and $c=\frac{1}{3}+\varepsilon$.

We apply Cauchy's residue theorem to the rectangular contour $\mathscr{C}$ joining the points $\frac{1}{4}-iT$, $\frac{1}{4}+iT$, $c+iT$ and $c-iT$. Now it is clear that

$$\mathrm{Res}_{s=\frac{1}{3}}\Big\{E_1\Big(s-\frac{1}{3}\Big)E_2\Big(s-\frac{1}{3}\Big)\frac{B^s}{s}\Big\}=\frac{E_2(0)}{48}B^{\frac{1}{3}}P(\log B),$$

for some monic polynomial $P\in\mathbb{R}[x]$ of degree 3. Define the difference

$$\mathscr{E}(B)=\sum_{n\leqslant B}\Delta(n)-\frac{E_2(0)}{48}B^{\frac{1}{3}}P(\log B).$$

Then it follows that

$$\mathscr{E}(B)\ll_\varepsilon\frac{B^{\frac{1}{3}+\varepsilon}}{T}+\Big(\int_{\frac{1}{4}-iT}^{\frac{1}{4}+iT}+\int_{\frac{1}{4}-iT}^{c-iT}+\int_{c+iT}^{\frac{1}{4}+iT}\Big)\Big|E_1\Big(s-\frac{1}{3}\Big)\frac{B^s}{s}\Big|\mathrm{d}s,$$

since $E_2(s-\frac{1}{3})$ is holomorphic and bounded on the half-plane $\Re e(s)\geqslant\frac{1}{4}$.

We proceed to estimate the contribution from the horizontal contours. Recall the well-known subconvexity bounds

$$\zeta(\sigma+it)\ll_\varepsilon\begin{cases}|t|^{\frac{1-\sigma}{3}+\varepsilon}, & \text{if }\sigma\in[\frac{1}{2},1],\\ |t|^{\frac{3-4\sigma}{6}+\varepsilon}, & \text{if }\sigma\in[0,\frac{1}{2}],\end{cases}$$

for any $|t|\geqslant 1$. A proof of these can be found in [117, Section II.3.4], for example. It therefore follows that

$$E_1\Big(\sigma-\frac{1}{3}+it\Big)\ll_\varepsilon|t|^{1-3\sigma+\varepsilon} \tag{5.21}$$

for any $\sigma\in[\frac{1}{4},\frac{1}{3})$ and any $|t|\geqslant 1$. We may now deduce that

$$\begin{aligned}\int_{\frac{1}{4}\pm iT}^{c\pm iT}\Big|E_1\Big(s-\frac{1}{3}\Big)\frac{B^s}{s}\Big|\mathrm{d}s&\ll_\varepsilon\int_{\frac{1}{4}}^{c}B^\sigma T^{-3\sigma+\varepsilon}\mathrm{d}\sigma\\&\ll_\varepsilon\frac{B^{\frac{1}{3}+\varepsilon}T^\varepsilon}{T}+\frac{B^{\frac{1}{4}}T^\varepsilon}{T^{\frac{3}{4}}}.\end{aligned} \tag{5.22}$$

Turning to the vertical contour, (5.21) gives

$$\begin{aligned}\int_{\frac{1}{4}-iT}^{\frac{1}{4}+iT}\Big|E_1\Big(s-\frac{1}{3}\Big)\frac{B^s}{s}\Big|\mathrm{d}s&\ll B^{\frac{1}{4}}\int_{-T}^{T}\frac{|E_1(-\frac{1}{12}+it)|}{1+|t|}\mathrm{d}t\\&\ll B^{\frac{1}{4}}\int_{-T}^{T}\frac{|t|^{\frac{1}{4}+\varepsilon}}{1+|t|}\mathrm{d}t\\&\ll B^{\frac{1}{4}}T^{\frac{1}{4}+\varepsilon}.\end{aligned}$$

Once combined with (5.22), we conclude that

$$\mathscr{E}(B) \ll_\varepsilon B^{\frac{1}{3}+\varepsilon}T^{-1+\varepsilon} + B^{\frac{1}{4}}T^{\frac{1}{4}+\varepsilon},$$

for any $T \geqslant 1$. Taking $T = B^{\frac{1}{15}}$ we obtain

$$\sum_{n \leqslant B} \Delta(n) = \frac{E_2(0)}{48} B^{\frac{1}{3}} P(\log B) + O_\varepsilon(B^{\frac{4}{15}+\varepsilon}), \tag{5.23}$$

for any $\varepsilon > 0$.

We are now ready to complete the proof of Theorem 2.4. For this it suffices to combine the latter estimate with partial summation in (5.17), before then applying Lemma 5.2. In this way we deduce that

$$\begin{aligned} N_U(B) &= 2T(B) + O(B\log B) \\ &= \frac{\sigma_\infty E_2(0)}{144} BQ(\log B) + O_\varepsilon(B^{1-\frac{1}{15}+\varepsilon}) + O(B\log B), \end{aligned}$$

for a further cubic monic polynomial $Q \in \mathbb{R}[x]$. Here $\sigma_\infty = 6\int_0^1 F_2(u)\mathrm{d}u$ is given by (2.22) and it follows from (5.20) that

$$E_2(0) = \prod_p \Big(1 - \frac{1}{p}\Big)^4 \Big(1 + \frac{4}{p} + \frac{1}{p^2}\Big).$$

This therefore completes the proof of (5.3), and so the proof of Theorem 2.4.

We have gone through the proof of (5.23) in considerable detail, using the tools of complex analysis. In practice, especially if one does not particularly care about obtaining an explicit error term, one can side-step this part of the argument by applying a standard *Tauberian theorem*. We leave it for the reader to check that the following result (proved in Narkiewicz [95], for example) can be readily adapted to handle the sum $\sum_{n \leqslant B} \Delta(n)$.

Lemma 5.5. *Recall the definition* (5.19) *of the Dirichlet series $H(s)$ and assume that $h(n) \geqslant 0$ for each $n \in \mathbb{N}$. Suppose that $H(s)$ is absolutely convergent for $\Re e(s) > \alpha > 0$ and that in its domain of convergence we have*

$$H(s) = \frac{F(s)}{(s-\alpha)^w} + G(s),$$

where $F(s), G(s)$ are holomorphic functions on the half-plane $\Re e(s) \geqslant \alpha$, with $F(\alpha) \neq 0$ and $w > 0$. Then we have

$$\sum_{n \leqslant x} h(n) \sim \frac{F(\alpha)}{\alpha\Gamma(w)} \cdot x^\alpha (\log x)^{w-1},$$

as $x \to \infty$.

Exercises for Chapter 5

Exercise 5.1. Let $b \geqslant a$ and $q \in \mathbb{N}$. Show that

$$\#\{n \in \mathbb{Z} \cap (a, b] : n \equiv n_0 \pmod q\} = \frac{b-a}{q} + O(1).$$

Exercise 5.2. Let $I \subset \mathbb{R}$ be an interval, let $a \in \mathbb{N}$ and let $f : \mathbb{R} \to \mathbb{R}_{\geqslant 0}$ be a function that is continuously differentiable on I and changes sign only finitely many times on I. Use Euler–Maclaurin summation (see (8.4)) to show that

$$\sum_{\substack{n \in \mathbb{Z} \cap I \\ \gcd(n,a)=1}} f(n) = \phi^*(a) \int_I f(t) \mathrm{d}t + O\big(2^{\omega(a)} \sup_{t \in I} |f(t)|\big).$$

Exercise 5.3. Show that

$$\sum_{\substack{d|n \\ \gcd(d,a)=1}} \frac{\mu(d)}{d} = \frac{\phi^*(n)}{\phi^*(\gcd(a,n))},$$

for any $a, n \in \mathbb{N}$.

Exercise 5.4. Let $\phi^\dagger(n) := \prod_{p|n}(1 + \frac{1}{p})$. Show that

$$\sum_{n \leqslant x} \phi^\dagger(n) = O(x).$$

Use this to show how the overall contribution from the error term in (5.15) can be improved from $O(B \log B)$ to $O(B)$.

Chapter 6

D_4 del Pezzo surface of degree 3

In this chapter we consider Manin's conjecture for the cubic surface

$$S_2 = \{x_1x_2(x_1+x_2) + x_4(x_1+x_2+x_3)^2 = 0\},$$

introduced in (2.14) According to Table 2.2 the surface contains exactly 6 lines, and one checks that the lines

$$x_i = x_4 = 0, \quad x_1+x_2 = x_j = 0, \quad x_i = x_1+x_2+x_3 = 0 \tag{6.1}$$

are all contained in S_2, for distinct indices $i \in \{1,2\}$ and $j \in \{3,4\}$. Let $U_2 \subset S_2$ be the open subset formed by deleting these lines from S_2. Our task is to estimate $N_{U_2}(B)$, with the goal of establishing Theorem 2.1. Our proof of this result will be in two stages. Firstly, we will give a completely self-contained account of the bound $N_{U_2}(B) = O_\varepsilon(B^{1+\varepsilon})$, before then indicating some of the extra technology needed to prove

$$N_{U_2}(B) \ll B(\log B)^6. \tag{6.2}$$

This will then establish Theorem 2.1.

In estimating $N_{U_2}(B)$ it will suffice to consider any surface that is obtained from S_2 via a unimodular transformation. In view of this we will make the change of variables

$$t_1 = x_1, \quad t_2 = x_2, \quad t_3 = x_1+x_2+x_3, \quad t_4 = -x_4,$$

which brings S_2 into the shape

$$t_1t_2(t_1+t_2) = t_3^2t_4, \tag{6.3}$$

and which we henceforth denote by S. The 6 lines on this surface take the shape

$$t_i = t_j = 0, \quad t_j = t_1+t_2 = 0,$$

for $i \in \{1,2\}$ and $j \in \{3,4\}$. If $U \subset S$ denotes the open subset formed by deleting these lines from the surface, then we have $t_3 t_4 = 0$ for any $[\mathbf{t}] \notin U$. It now follows that

$$N_U(B) = \frac{1}{2}\#\{\mathbf{t} \in \mathbb{Z}^4_{\text{prim}} : \text{ (6.3) holds, } |\mathbf{t}| \leqslant B, \ t_3 t_4 \neq 0\}.$$

There is a clear symmetry between solutions such that t_3 is positive and negative. Similarly, (6.3) is invariant under the transformation $t_1 = -z_1, t_2 = -z_2, t_3 = z_3$ and $t_4 = -z_4$. Thus we have

$$N_U(B) = 2\#\{\mathbf{t} \in \mathbb{Z}^4_{\text{prim}} : \text{ (6.3) holds, } |\mathbf{t}| \leqslant B, \ t_3, t_4 \geqslant 1\}, \tag{6.4}$$

for any $B \geqslant 1$.

6.1 Passage to the universal torsor

Let $\widetilde{S}$ denote the minimal desingularisation of the surface S. Derenthal [53] has calculated the universal torsor over $\widetilde{S}$, it being embedded in $\mathbb{A}^{10}$ by a single equation

$$s_1 u_1 y_1^2 + s_2 u_2 y_2^2 + s_3 u_3 y_3^2 = 0. \tag{6.5}$$

When it comes to counting integral points on the universal torsor, we will be led to consider the counting problem for rational points on plane curves of degree 1 and 2. We will need to do this uniformly with respect to the coefficients of the equations defining the curves for which our work in Chapter 4 will enter the picture.

Given $v \in \mathbb{R}$ and $\mathbf{s}, \mathbf{u}, \mathbf{y} \in \mathbb{R}^3$, define

$$\Psi(v, \mathbf{s}, \mathbf{u}, \mathbf{y}) := \max\left\{ \begin{array}{l} |s_1 s_2 s_3|, \ |u_1^2 u_2^2 u_3^2 v^3 y_1 y_2 y_3| \\ |s_1 u_1^2 u_2 u_3 v^2 y_1^2|, \ |s_2 u_1 u_2^2 u_3 v^2 y_2^2| \end{array} \right\}. \tag{6.6}$$

We are now ready to explicate the relation between (6.4) and counting integral points on (6.5). Our argument is analogous to that presented in Section 5.1, although the individual steps differ somewhat.

Lemma 6.1. *We have*

$$N_U(B) = 2\#\left\{ (v, \mathbf{s}, \mathbf{u}, \mathbf{y}) \in \mathbb{N}^4 \times \mathbb{Z}^3 \times \mathbb{N}^3 : \begin{array}{l} u_3 > 0, \ \Psi(v, \mathbf{s}, \mathbf{u}, \mathbf{y}) \leqslant B, \\ \text{(6.5) holds,} \\ |\mu(u_1 u_2 u_3)| = 1, \\ \gcd(s_1 s_2 s_3, u_1 u_2 u_3 v) = 1, \\ \gcd(y_i, y_j) = 1, \\ \gcd(y_i, s_j, s_k) = 1 \end{array} \right\}.$$

where i, j, k denote distinct elements from the set $\{1, 2, 3\}$.

Proof. Let $\mathbf{t} \in \mathbb{Z}^4$ be a vector such that (6.3) holds, with $t_3, t_4 \geqslant 1$. Write

$$\eta_{14} = \gcd(t_1, t_4), \quad \eta_{24} = \gcd(t_2, t_4\eta_{14}^{-1}), \quad \eta_{12} = \gcd(t_1\eta_{14}^{-1}, t_2\eta_{24}^{-1}).$$

Then $\eta_{12}, \eta_{14}, \eta_{24} \in \mathbb{N}$ and there exist $z_4 \in \mathbb{N}$ and $z_1, z_2 \in \mathbb{Z}$ such that

$$t_1 = \eta_{12}\eta_{14}z_1, \quad t_2 = \eta_{12}\eta_{24}z_2, \quad t_4 = \eta_{14}\eta_{24}z_4.$$

Moreover, it is not hard to deduce that

$$\gcd(\eta_{12}z_1, \eta_{24}z_4) = \gcd(\eta_{12}z_2, z_4) = \gcd(z_1, z_2) = 1,$$

and

$$\gcd(t_3, \eta_{14}, \eta_{12}\eta_{24}z_2) = 1.$$

Under this substitution the equation (6.3) becomes

$$\eta_{12}^3 z_1 z_2(\eta_{14}z_1 + \eta_{24}z_2) = t_3^2 z_4.$$

It follows that $\eta_{12}^3 \mid t_3^2$ in any given integer solution. Exercise 6.2 therefore implies that there exist $u, v, z_3 \in \mathbb{N}$ such that $|\mu(u)| = 1$ and

$$\eta_{12} = uv^2, \quad t_3 = u^2v^3z_3,$$

with

$$z_1z_2(\eta_{14}z_1 + \eta_{24}z_2) = uz_3^2z_4.$$

We proceed to consider the effect of the divisibility condition $z_1z_2 \mid uz_3^2$ that this equation entails.

Recall that $\gcd(z_1, z_2) = \gcd(z_1, z_4) = \gcd(z_2, z_4) = 1$. Since $z_1z_2 \mid uz_3^2$, there must exist $u_1, u_2, u_3, w_1, w_2, w_3 \in \mathbb{Z}$ such that $w_1, w_2, w_3, u_3 > 0$ and

$$u = u_1u_2u_3, \quad z_1 = u_1w_1^2, \quad z_2 = u_2w_2^2, \quad z_3 = w_1w_2w_3.$$

Here we have used the fact that if p is a prime such that $p \nmid u$ and $p \mid z_1z_2$, then p must divide z_1 or z_2 to even order. Under these substitutions our equation becomes

$$\eta_{14}u_1w_1^2 + \eta_{24}u_2w_2^2 = u_3w_3^2z_4.$$

Moreover, we will have the corresponding coprimality conditions

$$\gcd(u_1u_2u_3vw_1, \eta_{24}z_4) = \gcd(u_1u_2u_3vw_2, z_4) = \gcd(u_1w_1, u_2w_2) = 1, \tag{6.7}$$

and

$$|\mu(u_1u_2u_3)| = 1, \quad \gcd(u_1^2u_2^2u_3^2v^3w_1w_2w_3, \eta_{14}, \eta_{24}u_1u_2^2u_3v^2w_2^2) = 1. \tag{6.8}$$

We now set $\mathbf{s} = (\eta_{14}, \eta_{24}, z_4)$ and $\mathbf{y} = \mathbf{w}$, and replace (u_1, u_2, u_3) by $(-u_1, -u_2, u_3)$. Tracing through our argument, one sees that we have made the transformation

$$\begin{cases} t_1 = -s_1 u_1^2 u_2 u_3 v^2 y_1^2, \\ t_2 = -s_2 u_1 u_2^2 u_3 v^2 y_2^2, \\ t_3 = u_1^2 u_2^2 u_3^2 v^3 y_1 y_2 y_3, \\ t_4 = s_1 s_2 s_3. \end{cases}$$

In particular, it is clear that the height condition $|\mathbf{x}| \leqslant B$ is equivalent to $\Psi(v, \mathbf{s}, \mathbf{u}, \mathbf{y}) \leqslant B$, in the notation of (6.6). We now observe that under this transformation the equation (6.3) becomes (6.5), and the coprimality relations (6.7) and (6.8) can be rewritten

$$\gcd(s_2 s_3, u_1 u_2 u_3 v y_1) = \gcd(s_3, u_1 u_2 u_3 v y_2) = \gcd(u_1 y_1, u_2 y_2) = 1,$$

and

$$|\mu(u_1 u_2 u_3)| = 1, \quad \gcd(s_1, u_1 u_2 u_3 v y_2 \gcd(y_3, s_2)) = 1.$$

We can combine these relations with (6.5) to simplify them still further. In fact, once combined with (6.5), we claim that they are equivalent to the conditions appearing in the statement of the lemma. To establish the forward implication, it suffices to show that $\gcd(y_1, y_3) = \gcd(y_2, y_3) = 1$, the remaining conditions being immediate. But these two conditions follow on combining (6.5) with the fact that $\gcd(y_1, s_2 u_2 y_2) = 1$. To see the reverse implication, the conditions are all immediate apart from

$$\gcd(y_1, s_2 s_3) = \gcd(y_2, s_1 s_3) = \gcd(u_1, y_2) = \gcd(u_2, y_1) = 1.$$

But each of these is an easy consequence of the assumed coprimality relations, and (6.5).

Finally, we leave it as an exercise to the reader to check that each $(v, \mathbf{s}, \mathbf{u}, \mathbf{y})$ counted in the right-hand side of Lemma 6.1 produces a primitive solution of (6.3) with $t_3, t_4 \geqslant 1$. This completes the proof of Lemma 6.1. □

In what follows let us write i for a generic element of the set $\{1, 2, 3\}$. Fix a choice of $v \in \mathbb{N}$ and $S_i, U_i, Y_i > 0$, and write

$$\mathcal{N} = \mathcal{N}_v(\mathbf{S}; \mathbf{U}; \mathbf{Y}) \tag{6.9}$$

for the total contribution to $N_U(B)$ in Lemma 6.1 from $\mathbf{s}, \mathbf{u}, \mathbf{y}$ contained in the intervals

$$\frac{S_i}{2} < s_i \leqslant S_i, \quad \frac{U_i}{2} < |u_i| \leqslant U_i, \quad \frac{Y_i}{2} < y_i \leqslant Y_i. \tag{6.10}$$

Write

$$S = S_1 S_2 S_3, \quad U = U_1 U_2 U_3, \quad Y = Y_1 Y_2 Y_3.$$

If $\mathscr{N} = 0$, there is nothing to prove, and so we assume henceforth that the ranges in (6.10) produce a non-zero value of $\mathscr{N}$. In particular we must have $S_i, U_i, Y_i \gg 1$ and

$$S \ll B \quad U^2Y \ll \frac{B}{v^3}, \quad S_iUU_iY_i^2 \ll \frac{B}{v^2}. \tag{6.11}$$

As indicated above we will provide two upper bounds for $N_U(B)$ in this chapter. By ignoring some of the technical machinery needed to get better bounds it is hoped that the overall methodology will be brought into focus. Later we will indicate how the expected upper bound can be retrieved with a little more work.

6.2 A crude upper bound

The object of this section is merely to establish linear growth, without worrying about the factor involving $\log B$ that we expect to see. Note that (6.11) forces the inequalities $S_i, U_i, Y_i \ll B$. We proceed to establish the following upper bound.

Lemma 6.2. *We have*

$$N_U(B) \ll (\log B)^9 \sum_{v \leqslant B^{\frac{1}{3}}} \max_{S_i,U_i,Y_i>0} \mathscr{N}_v(\mathbf{S};\mathbf{U};\mathbf{Y}),$$

where the maximum is over $S_i, U_i, Y_i > 0$ such that (6.11) *holds.*

Proof. Our starting point is Lemma 6.1. It follows from (6.6) that $v \leqslant B^{\frac{1}{3}}$ for any $(v, \mathbf{s}, \mathbf{u}, \mathbf{y})$ that contributes to the right-hand side. Let us fix a choice of $v \in \mathbb{N}$ such that $v \leqslant B^{\frac{1}{3}}$, and cover the ranges for $\mathbf{s}, \mathbf{u}, \mathbf{y}$ with dyadic intervals. Thus for fixed integers $\sigma_i, \nu_i, \eta_i \geqslant 0$, we write

$$S_i = 2^{\sigma_i}, \quad U_i = 2^{\nu_i}, \quad Y_i = 2^{\eta_i},$$

and consider the contribution from $\mathbf{s}, \mathbf{u}, \mathbf{y}$ in the range (6.10). But this is just $\mathscr{N} = \mathscr{N}_v(\mathbf{S};\mathbf{U};\mathbf{Y})$. Now we have already seen that $\mathscr{N} = 0$ unless (6.11) holds. Finally, since each S_i, U_i, Y_i is $O(B)$, it follows that the number of dyadic intervals needed is $O((\log B)^9)$. This completes the proof of the lemma. □

We may now restrict our attention to bounding $\mathscr{N}_v(\mathbf{S};\mathbf{U};\mathbf{Y})$ for fixed values of $S_i, U_i, Y_i \gg 1$ such that (6.11) holds, and fixed $v \leqslant B^{\frac{1}{3}}$. In the arguments that follow it will be necessary to focus attention on primitive $\mathbf{s} \in \mathbb{N}^3$. To enable this we draw out possible common factors between s_1, s_2, s_3, obtaining

$$\mathscr{N}_v(\mathbf{S};\mathbf{U};\mathbf{Y}) = \sum_{k=1}^{\infty} \mathscr{N}_v^*(k^{-1}\mathbf{S};\mathbf{U};\mathbf{Y}), \tag{6.12}$$

where $\mathscr{N}_v^*(\mathbf{S};\mathbf{U};\mathbf{Y})$ is defined as for $\mathscr{N}_v(\mathbf{S};\mathbf{U};\mathbf{Y})$ but with the extra condition that $\gcd(s_1, s_2, s_3) = 1$. Let us write $S_i' = k^{-1}S_i$ and $\mathbf{S}' = k^{-1}\mathbf{S}$.

Recall the equation (6.5) that we must count solutions to, which it will be convenient to denote by $\mathscr{T}$, and which we will think of as defining a variety in $\mathbb{P}^2 \times \mathbb{P}^2 \times \mathbb{P}^2$, with homogeneous coordinates $\mathbf{s}, \mathbf{u}, \mathbf{y}$. The key idea will be to count points on the fibres of projections $\pi : \mathscr{T} \to \mathbb{P}^2 \times \mathbb{P}^2$. This amounts to fixing six of the variables and estimating the number of points on the resulting plane curve. Since this family of curves will vary with B, so it is vital to apply bounds that are uniform in the coefficients of the defining equation.

Let us begin by fixing the variables $\mathbf{u}, \mathbf{y}$, and estimating the corresponding number of vectors $\mathbf{s}$. Now it follows from the coprimality conditions in Lemma 6.1 that

$$\gcd(u_1y_1^2, u_2y_2^2, u_3y_3^2) = 1.$$

For fixed $\mathbf{u}, \mathbf{y}$, (6.5) defines a line in $\mathbb{P}^2$. We clearly have

$$\mathscr{N}_v^*(\mathbf{S}'; \mathbf{U}; \mathbf{Y}) \leqslant \sum_{\mathbf{u},\mathbf{y}} M_1(\mathbf{a}; \mathbf{S}'),$$

in the notation of (4.16), with $a_i = u_iy_i^2$. Since $\mathbf{a}$ is primitive, it therefore follows from Lemma 4.8 that

$$\mathscr{N}_v^*(\mathbf{S}'; \mathbf{U}; \mathbf{Y}) \ll \sum_{\mathbf{u},\mathbf{y}} \Big(1 + \frac{S}{k^2 \max S_iU_iY_i^2}\Big) \ll UY + k^{-2}S^{\frac{2}{3}}U^{\frac{2}{3}}Y^{\frac{1}{3}}.$$

Here we have used the trivial lower bound $\max\{a, b, c\} \geqslant (abc)^{\frac{1}{3}}$, valid for any $a, b, c \geqslant 0$. Using (6.11) we conclude that

$$\mathscr{N}_v^*(\mathbf{S}'; \mathbf{U}; \mathbf{Y}) \ll UY + \frac{B}{k^2v}. \tag{6.13}$$

The second term here will be satisfactory from our point of view, but the first is disastrous, since we will run into trouble when it comes to summing over k in (6.12).

It turns out that an altogether different bound is required to handle the contribution from really small values of $\mathbf{S}'$. For this we will fix values of $\mathbf{s}, \mathbf{u}$ in (6.5), and count points on the resulting family of conics. First we need to record the coprimality relation

$$\gcd(s_iu_i, s_ju_j) = 1,$$

which we claim holds for any of the vectors $\mathbf{s}, \mathbf{u}, \mathbf{y}$ in which we are interested. But this follows on noting that $\gcd(s_i, s_j) = 1$ for any vector $\mathbf{s} \in \mathbb{Z}^3_{\text{prim}}$ such that (6.5) holds and $\gcd(s_i, u_1u_2u_3) = \gcd(y_i, s_j, s_k) = 1$. We now have

$$\mathscr{N}_v^*(\mathbf{S}'; \mathbf{U}; \mathbf{Y}) \leqslant \sum_{\mathbf{s},\mathbf{u}} M_2(\mathbf{a}; \mathbf{Y}),$$

in the notation of (4.16), with $a_i = s_i u_i$. In particular we have that a_i is non-zero and $\gcd(a_i, a_j) = 1$ in the statement of Lemma 4.10, whence

$$\mathscr{N}_v^*(\mathbf{S}'; \mathbf{U}; \mathbf{Y}) \ll \sum_{\mathbf{s},\mathbf{u}} \Big(1 + \frac{kY^{\frac{1}{3}}}{S^{\frac{1}{3}}U^{\frac{1}{3}}}\Big) 2^{\omega(s_1 s_2 s_3 u_1 u_2 u_3)}.$$

In view of the bounds $S, U \ll B$, we clearly have

$$2^{\omega(s_1 s_2 s_3 u_1 u_2 u_3)} \ll_\varepsilon (s_1 s_2 s_3 u_1 u_2 u_3)^\varepsilon \ll_\varepsilon (SU)^\varepsilon \ll_\varepsilon B^{2\varepsilon},$$

for any $\varepsilon > 0$. Once inserted into our bound for $\mathscr{N}_v^*(\mathbf{S}'; \mathbf{U}; \mathbf{Y})$, and combined with (6.11), we deduce that

$$\mathscr{N}_v^*(\mathbf{S}'; \mathbf{U}; \mathbf{Y}) \ll_\varepsilon B^{2\varepsilon}\Big(\frac{SU}{k^3} + \frac{S^{\frac{2}{3}}U^{\frac{2}{3}}Y^{\frac{1}{3}}}{k^2}\Big) \ll_\varepsilon \frac{SUB^{2\varepsilon}}{k^3} + \frac{B^{1+2\varepsilon}}{k^2 v}. \tag{6.14}$$

Here the second term will provide a satisfactory contribution.

Note that

$$\min\Big\{\frac{SU}{k^3}, UY\Big\} \leqslant \frac{U\sqrt{SY}}{k^{\frac{3}{2}}} \ll \frac{B}{k^{\frac{3}{2}}v^{\frac{3}{2}}},$$

by (6.11). It therefore follows from (6.13) and (6.14) that

$$\mathscr{N}_v^*(\mathbf{S}'; \mathbf{U}; \mathbf{Y}) \ll_\varepsilon \frac{B^{1+2\varepsilon}}{k^{\frac{3}{2}}v}.$$

Once inserted into (6.12), and then into the statement of Lemma 6.2, we may conclude that

$$N_U(B) \ll_\varepsilon (\log B)^9 \sum_{v \leqslant B^{\frac{1}{3}}} \sum_{k=1}^{\infty} \frac{B^{1+2\varepsilon}}{k^2 v} \ll_\varepsilon B^{1+2\varepsilon}(\log B)^{10} \ll_\varepsilon B^{1+3\varepsilon}.$$

On redefining the choice of parameter $\varepsilon > 0$, we have therefore established the bound $N_U(B) \ll_\varepsilon B^{1+\varepsilon}$.

The reader will note that there have been many places in our argument where we have been wasteful. The most damaging has been in our use of the trivial bound $2^{\omega(n)} = O_\varepsilon(n^\varepsilon)$, in the deduction of (6.14). Using the fact that $2^{\omega(n)}$ has average order $\frac{1}{\zeta(2)} \log n$, it is not particularly difficult to replace the B^ε by a power of $\log B$. In the following section we show that one can replace B^ε by $(\log B)^6$.

6.3 A better upper bound

Crucial to the work in the preceding section was an investigation of the density of integer solutions to the equation (6.5). It is in our treatment of this equation that we will hope to gain some saving.

Let us put the problem on a more general footing. For any $\mathbf{A}, \mathbf{B}, \mathbf{C} \in \mathbb{R}^3_{\geqslant 1}$, let $\mathscr{M}(\mathbf{A}, \mathbf{B}, \mathbf{C})$ denote the number of $\mathbf{a}, \mathbf{b}, \mathbf{c} \in Z^3_*$ such that

$$a_1 b_1 c_1^2 + a_2 b_2 c_2^2 + a_3 b_3 c_3^2 = 0 \tag{6.15}$$

and

$$|a_i| \leqslant A_i, \quad |b_i| \leqslant B_i, \quad |c_i| \leqslant C_i,$$

with $|\mu(a_1 a_2 a_3)| = 1$ and

$$\gcd(a_i, c_j) = \gcd(c_i, c_j) = \gcd(a_i, b_j, b_k) = 1. \tag{6.16}$$

Here, we recall that Z^3_* denotes the set of primitive vectors in $\mathbb{Z}^3$ with all components non-zero. It will be convenient to set

$$A = A_1 A_2 A_3, \quad B = B_1 B_2 B_3, \quad C = C_1 C_2 C_3.$$

Arguing exactly as in the previous section, it is not difficult to deduce from Lemmas 4.8 and 4.10 that

$$\mathscr{M}(\mathbf{A}, \mathbf{B}, \mathbf{C}) \ll_\varepsilon A \min\{C, A^\varepsilon B^{1+\varepsilon}\} + A^{\frac{2}{3}+\varepsilon} B^{\frac{2}{3}+\varepsilon} C^{\frac{1}{3}}$$

for any $\varepsilon > 0$, whence

$$\mathscr{M}(\mathbf{A}, \mathbf{B}, \mathbf{C}) \ll_\varepsilon A^{\frac{2}{3}+\varepsilon} B^{\frac{2}{3}+\varepsilon} C^{\frac{1}{3}} + A^{1+\varepsilon} B^{\frac{1}{2}+\varepsilon} C^{\frac{1}{2}}. \tag{6.17}$$

By working a little harder, we would like to replace the terms A^ε, B^ε by something rather smaller.

The main problem to be faced emerges in the application of Lemma 4.10, which gives

$$\mathscr{M}(\mathbf{A}, \mathbf{B}, \mathbf{C}) \ll \sum_{\mathbf{a}, \mathbf{b}} \Big(1 + \frac{C^{\frac{1}{3}}}{|a_1 a_2 a_3 b_1 b_2 b_3|^{\frac{1}{3}}}\Big) \tau(a_1 a_2 a_3 b_1 b_2 b_3),$$

on taking $2^{\omega(n)} \leqslant \tau(n)$. Rather than using the trivial bound $\tau(n) = O_\varepsilon(n^\varepsilon)$, it is fairly straightforward to show that $A^\varepsilon B^\varepsilon$ can be replaced by $(\log A)^{\frac{3}{2}} (\log B)^3$ in (6.17). However this would still not be enough to deduce the upper bound for $N_U(B)$ that we would like.

Let us simplify matters by considering only the contribution

$$\mathscr{S}(\mathbf{A}, \mathbf{B}) = \sum_{\mathbf{a}, \mathbf{b}} \tau(a_1 a_2 a_3 b_1 b_2 b_3),$$

in the above estimate for $\mathscr{M}(\mathbf{A}, \mathbf{B}, \mathbf{C})$. Then $\mathscr{S}(\mathbf{A}, \mathbf{B})$ has exact order of magnitude

$$AB \prod_{i=1}^{3} (\log A_i)(\log B_i),$$

so how can we hope to do better than this? The crucial observation comes in noting that we are only interested in summing over values of $\mathbf{a}, \mathbf{b}$ for which the corresponding conic (6.15) has a non-zero solution $\mathbf{c} \in \mathbb{Z}^3$, with $\gcd(c_i, c_j) = 1$. If we denote this finer quantity by $\mathscr{S}^*(\mathbf{A}, \mathbf{B})$, then it is actually possible to show that

$$\mathscr{S}^*(\mathbf{A}, \mathbf{B}) \ll AB.$$

This is essentially established in [22, Proposition 1] and is a consequence of the fact that a random plane conic does not possess a rational point. This should be compared with the work of Serre [111]. Using the large sieve inequality, Serre has shown

$$\#\{\mathbf{y} \in \mathbb{Z}^3 : |\mathbf{y}| \leqslant Y, \ (-y_1y_3, -y_2y_3)_{\mathbb{Q}} = 1\} \ll \frac{Y^3}{(\log Y)^{\frac{3}{2}}},$$

where

$$(a,b)_{\mathbb{Q}} = \begin{cases} 1, & \text{if } ax^2 + by^2 = z^2 \text{ has a solution } (x,y,z) \neq \mathbf{0} \text{ in } \mathbb{Q}^3, \\ -1, & \text{otherwise,} \end{cases}$$

denotes the *Hilbert symbol.* Guo [63] has established an asymptotic formula for the corresponding quantity in which one counts only odd values of y_1, y_2, y_3 such that the product $y_1y_2y_3$ is square-free. Crucial to both of these works is the fact $(a,b)_{\mathbb{Q}} = 1$ if and only if $(a,b)_{\mathbb{R}} = 1$ and $(a,b)_{\mathbb{Q}_p} = 1$ for every prime p, where for an arbitrary field K the Hilbert symbol $(a,b)_K$ is defined analogously as for $K = \mathbb{Q}$. This is the *Hasse–Minkowski theorem* and ensures that quadratic forms satisfy the Hasse principle.

In addition to considering the density of solutions to diagonal quadratic equations, as in the previous section, we therefore need to consider how often such an equation has at least one non-trivial integer solution in order to derive sufficiently sharp bounds. The outcome of this investigation is the following result, which is established in [22, Proposition 2].

Lemma 6.3. *For any* $\varepsilon > 0$, *we have*

$$\mathscr{M}(\mathbf{A}, \mathbf{B}, \mathbf{C}) \ll_\varepsilon A^{\frac{2}{3}} B^{\frac{2}{3}} C^{\frac{1}{3}} + \sigma\tau A B^{\frac{1}{2}} C^{\frac{1}{2}},$$

where

$$\sigma = 1 + \frac{\min\{A, B\}^\varepsilon}{\min\{B_i B_j\}^{\frac{1}{16}}}, \quad \tau = 1 + \frac{\log B}{\min\{B_i B_j\}^{\frac{1}{16}}}.$$

It is clear that this constitutes a substantial sharpening over our earlier estimate (6.17) for $\mathscr{M}(\mathbf{A}, \mathbf{B}, \mathbf{C})$. Nonetheless this is still not enough on its own, and we will need an alternative estimate when B_1, B_2, B_3 have particularly awkward sizes. The following result is rather easy to establish.

Lemma 6.4. *We have*

$$\mathscr{M}(\mathbf{A}, \mathbf{B}, \mathbf{C}) \ll A B_i B_j (C_k + C_i C_j A_k^{-1}) (\log AC)^2,$$

for any permutation $\{i, j, k\}$ *of the set* $\{1, 2, 3\}$.

Proof. For fixed integers a, b, q, let $\rho(q; a, b)$ denote the number of solutions to the congruence $at^2 + b \equiv 0 \pmod q$. We then have

$$\rho(q; a, b) \leqslant \sum_{d|q} |\mu(d)| \Big(\frac{-ab}{d}\Big),$$

by Exercise 6.4 It will clearly suffice to establish Lemma 6.4 in the case $(i, j, k) = (1, 2, 3)$, say. Now it follows from (6.15) that for given a_i, b_1, b_2, c_3, and each corresponding solution t of the congruence

$$a_1 b_1 t^2 + a_2 b_2 \equiv 0 \pmod{a_3 c_3^2},$$

we must have $c_1 \equiv tc_2 \pmod{a_3 c_3^2}$ in any solution to be counted. This gives rise to an equation of the form $\mathbf{h}.\mathbf{w} = 0$, with $\mathbf{h} = (1, -t, a_3 c_3^2)$ and $\mathbf{w} = (c_1, c_2, k)$. Upon recalling that $\gcd(c_1, c_2) = 1$ from (6.16), an application of Lemma 4.8 therefore yields the bound

$$\ll \rho(a_3 c_3^2; a_1 b_2, a_2 b_2)\Big(1 + \frac{C_1 C_2}{|a_3 c_3^2|}\Big),$$

for the number of possible b_3, c_1, c_2 given fixed choices of a_i, b_1, b_2 and c_3. It now follows from our upper bound for ρ that

$$\begin{aligned}\mathscr{M}(\mathbf{A}, \mathbf{B}, \mathbf{C}) &\ll \sum_{a_i, b_1, b_2, c_3} \rho(a_3 c_3^2; a_1 b_2, a_2 b_2)\Big(1 + \frac{C_1 C_2}{|a_3 c_3^2|}\Big) \\ &\ll \sum_{a_i, b_1, b_2, c_3} \sum_{d | a_3 c_3} |\mu(d)| \Big(\frac{-a_1 a_2 b_1 b_2}{d}\Big)\Big(1 + \frac{C_1 C_2}{|a_3 c_3^2|}\Big) \\ &\ll \sum_{a_i, b_1, b_2, c_3} \tau(a_3)\tau(c_3) + C_1 C_2 \sum_{a_i, b_1, b_2, c_3} \frac{\tau(a_3)\tau(c_3)}{|a_3 c_3^2|}.\end{aligned}$$

A simple application of partial summation now reveals that

$$\mathscr{M}(\mathbf{A}, \mathbf{B}, \mathbf{C}) \ll \big(AB_1 B_2 C_3 + A_1 A_2 B_1 B_2 C_1 C_2\big)(\log AC)^2,$$

as required to complete the proof of Lemma 6.4. □

We are now ready to combine Lemmas 6.3 and 6.4 to get a sharper upper bound for $N_U(B)$. Taking Lemma 6.1 as our starting point we need to bound the quantity $\mathscr{N} = \mathscr{N}_v(\mathbf{S}; \mathbf{U}; \mathbf{Y})$ defined in (6.9), for fixed choices of $v \in \mathbb{N}$ and $S_i, U_i, Y_i > 0$. As previously we will need to extract common factors from s_1, s_2, s_3, leading to the equality (6.12). Writing $S_i' = k^{-1} S_i$ and $\mathbf{S}' = k^{-1}\mathbf{S}$, as before, it is a simple matter to check that we have

$$\mathscr{N}_v^*(\mathbf{S}'; \mathbf{U}; \mathbf{Y}) \leqslant \mathscr{M}(\mathbf{U}, \mathbf{S}', \mathbf{Y}),$$

with $(\mathbf{a}, \mathbf{b}, \mathbf{c}) = (\mathbf{u}, \mathbf{s}, \mathbf{y})$. Indeed we plainly have

$$\gcd(u_i, y_j) = \gcd(y_i, y_j) = 1, \quad |\mu(u_1 u_2 u_3)| = \gcd(u_i, s_j, s_k) = 1,$$

and $\mathbf{u}, \mathbf{s}, \mathbf{y} \in Z_*^3$, for any vectors counted by $\mathscr{N}_v^*(\mathbf{S}'; \mathbf{U}; \mathbf{Y})$, as required for $\mathscr{M}(\mathbf{U}, \mathbf{S}', \mathbf{Y})$. It now follows from (6.12) and Lemma 6.3 that

$$\begin{aligned}\mathscr{N}_v(\mathbf{S}; \mathbf{U}; \mathbf{Y}) &\ll_\varepsilon \sum_{k=1}^{\infty} \Big(\frac{U^{\frac{2}{3}} S^{\frac{2}{3}} Y^{\frac{1}{3}}}{k^2} + k^{\frac{1}{8}} \sigma \tau \frac{U S^{\frac{1}{2}} Y^{\frac{1}{2}}}{k^{\frac{3}{2}}} \Big) \\ &\ll_\varepsilon U^{\frac{2}{3}} S^{\frac{2}{3}} Y^{\frac{1}{3}} + \sigma \tau U S^{\frac{1}{2}} Y^{\frac{1}{2}},\end{aligned}$$

for any $\varepsilon > 0$, where

$$\sigma = 1 + \frac{\min\{S, U\}^\varepsilon}{\min\{S_i S_j\}^{\frac{1}{16}}}, \quad \tau = 1 + \frac{\log B}{\min\{S_i S_j\}^{\frac{1}{16}}}.$$

In order to obtain our final estimate for $N_U(B)$ we need to sum this bound over all positive integers $v \leqslant B^{\frac{1}{3}}$, as in Lemma 6.2, and over all possible dyadic intervals for S_i, U_i, Y_i, subject to (6.11).

Suppose for the moment that we want to sum over all possible dyadic intervals $X \leqslant |x| < 2X$, for which $|x| \leqslant \mathscr{X}$. Then in deducing Lemma 6.2 we employed the fact that it is enough to consider $O(\log \mathscr{X})$ choices for X. In the present investigation we will be more efficient and take advantage of the estimates

$$\sum_X X^\delta \ll_\delta \begin{cases} 1, & \text{if } \delta < 0, \\ \mathscr{X}^\delta, & \text{if } \delta > 0, \end{cases}$$

where the sum is over dyadic intervals for $X \leqslant \mathscr{X}$. We will make frequent use of these bounds without further mention.

Returning to our estimate for $\mathscr{N}_v(\mathbf{S}; \mathbf{U}; \mathbf{Y})$, we may conclude from the bound $Y_i \leqslant B^{\frac{1}{2}} (v^2 S_i U U_i)^{-\frac{1}{2}}$ in (6.11) that

$$\begin{aligned}N_U(B) &\ll_\varepsilon \sum_{v \leqslant B^{\frac{1}{3}}} \sum_{S_i, U_i, Y_i} \Big(U^{\frac{2}{3}} S^{\frac{2}{3}} Y^{\frac{1}{3}} + \sigma \tau U S^{\frac{1}{2}} Y^{\frac{1}{2}} \Big) \\ &\ll_\varepsilon B^{\frac{1}{2}} \sum_{v \leqslant B^{\frac{1}{3}}} \sum_{S_i, U_i} \frac{S^{\frac{1}{2}}}{v} + \sum_{v \leqslant B^{\frac{1}{3}}} \sum_{S_i, U_i, Y_i} \sigma \tau U S^{\frac{1}{2}} Y^{\frac{1}{2}} \\ &\ll_\varepsilon B (\log B)^6 + \sum_{v \leqslant B^{\frac{1}{3}}} \sum_{S_i, U_i, Y_i} \sigma \tau U S^{\frac{1}{2}} Y^{\frac{1}{2}}.\end{aligned}$$

The first term on the right-hand side is clearly satisfactory, and it remains to deal with the second term, which we denote by R for convenience. We would like to show that $R \ll B(\log B)^6$, in order to complete the proof of (6.2).

Suppose without loss of generality that $S_1 \leqslant S_2 \leqslant S_3$, so that in particular $\min\{S_i S_j\} = S_1 S_2$ in σ and τ. If there is a constant $A > 0$ such that $S_3 \leqslant (S_1 S_2)^A$, then it follows that

$$\sigma \leqslant 1 + (S_1 S_2)^{\varepsilon - \frac{1}{16}} S_3^\varepsilon \leqslant 1 + (S_1 S_2)^{(1+A)\varepsilon - \frac{1}{16}} \ll 1,$$

provided that ε is sufficiently small. Taking $\tau \ll \log B$, we may then argue as above to conclude that there is a contribution of $O(B(\log B)^6)$ to R from this case. Suppose now that there exists $A' > 0$ such that $U \leqslant (S_1S_2)^{A'}$. Then we have $\sigma \ll 1$ and $\tau \ll \log B$, so that there is a contribution of $O(B(\log B)^6)$ to R in this case too.

Finally it remains to consider the contribution to $N_U(B)$ from S_i, U_i, Y_i such that

$$S_1S_2 \leqslant \min\{S_3, U\}^\delta, \tag{6.18}$$

for some small value of $\delta > 0$, with $S_1 \leqslant S_2 \leqslant S_3$. Let us denote this contribution N_0, say. To estimate N_0 we will return to the task of estimating $\mathscr{N}_v(\mathbf{S};\mathbf{U};\mathbf{Y})$ for fixed v, S_i, U_i, Y_i, but this time apply Lemma 6.4 with $(i,j,k) = (1,2,3)$. This gives

$$\mathscr{N}_v(\mathbf{S};\mathbf{U};\mathbf{Y}) \ll (\log B)^2\big(US_1S_2Y_3 + U_1U_2S_1S_2Y_1Y_2\big).$$

We must now sum over dyadic intervals for S_i, U_i, Y_i. Thus it follows from the bound $Y_i \leqslant B^{\frac{1}{2}}(v^2S_iUU_i)^{-\frac{1}{2}}$ in (6.11) that

$$\begin{aligned}
N_0 &\ll (\log B)^2 \sum_{v \leqslant B^{\frac{1}{3}}} \sum_{S_i,U_i,Y_i} (US_1S_2Y_3 + U_1U_2S_1S_2Y_1Y_2)\\
&\ll (\log B)^2 \sum_{v \leqslant B^{\frac{1}{3}}} \Big(\sum_{\substack{S_i,U_i\\ Y_1,Y_2}} \frac{B^{\frac{1}{2}}U^{\frac{1}{2}}S_1S_2}{vS_3^{\frac{1}{2}}U_3^{\frac{1}{2}}} + \sum_{S_i,U_i,Y_3} \frac{B(S_1S_2U_1U_2)^{\frac{1}{2}}}{v^2U}\Big).
\end{aligned}$$

Since $U^2 \ll B(v^3Y_1Y_2)^{-1}$ in (6.11), and $S_1S_2 \ll S_3^\delta$ by (6.18), we therefore deduce that the overall contribution from the first inner sum is

$$\begin{aligned}
&\ll (\log B)^2 \sum_{v \leqslant B^{\frac{1}{3}}} \sum_{\substack{S_i,Y_1,Y_2\\ U_2,U_3}} \frac{B^{\frac{3}{4}}S_1S_2}{v^{\frac{7}{4}}S_3^{\frac{1}{2}}U_3^{\frac{1}{2}}Y_1^{\frac{1}{4}}Y_2^{\frac{1}{4}}}\\
&\ll \sum_{\substack{S_2,S_3,Y_1,Y_2\\ U_2,U_3}} \frac{B^{\frac{3}{4}}(\log B)^2}{S_3^{\frac{1}{2}-\delta}U_3^{\frac{1}{2}}Y_1^{\frac{1}{4}}Y_2^{\frac{1}{4}}}\\
&\ll B.
\end{aligned}$$

Turning to the contribution from the second inner sum, we deduce from a second application of (6.18) that

$$\begin{aligned}
N_0 &\ll B + (\log B)^2 \sum_{S_i,U_i,Y_3} \frac{B(S_1S_2U_1U_2)^{\frac{1}{2}}}{U}\\
&\ll B + (\log B)^2 \sum_{S_2,S_3,U_i,Y_3} \frac{B}{U^{\frac{1-\delta}{2}}}\\
&\ll B(\log B)^5.
\end{aligned}$$

Once combined with our earlier work, this therefore concludes the proof of (6.2), and so the proof of Theorem 2.1.

Exercises for Chapter 6

Exercise 6.1. Check that (6.1) are all of the lines contained in S_2.

Exercise 6.2. Let $a, b \in \mathbb{N}$. Show that $a \mid b^2$ if and only if $a = uv^2$ for $u, v \in \mathbb{N}$ such that u is square-free and $uv \mid b$.

Exercise 6.3. Use the large sieve inequality to show that there are $o(H^2)$ positive integers $a, b \leqslant H$ such that the conic $ax^2 + by^2 = z^2$ has a rational point.

Exercise 6.4. For given $a, b, q \in \mathbb{Z}$ such that $q > 0$ show that

$$\#\{t \pmod q : at^2 + b \equiv 0 \pmod q\} \leqslant \sum_{d \mid q} |\mu(d)| \Big(\frac{-ab}{d}\Big),$$

where $(\frac{\cdot}{d})$ is the Jacobi symbol.

Exercise 6.5. By mimicking the argument in [22, Section 5], establish the lower bound $N_{U_2}(B) \gg B(\log B)^6$ to complement (6.2).

Chapter 7

Siegel's lemma and non-singular surfaces

In Chapter 4 we met a number of upper bounds for the density of rational points on open subsets of surfaces in $\mathbb{P}^3$. In particular, if $X \subset \mathbb{P}^3$ is a non-singular surface of degree $d \geqslant 3$ and $U \subseteq X$ is the open subset formed by deleting the lines, then we saw in Theorem 4.6 that

$$N_U(B) = O_{d,\varepsilon}(B^{1+\varepsilon} + B^{\frac{3}{\sqrt{d}}+\frac{2}{d-1}+\varepsilon}), \tag{7.1}$$

for any $\varepsilon > 0$. One easily gets an improvement of this by combining Lemma 4.2 with Theorem 4.7. Both of these results rely upon covering the points counted by $N_U(B)$ by a bounded number of hypersurface sections. This therefore reduces the dimension of the problem, leaving one to count points on all of the curves that are produced in this way.

One notes that (7.1) is worse than the exponent predicted by Conjecture 3.4 when $n = 4$ and $d = 3$. The purpose of this chapter is to discuss a further means of covering the rational points of low height on a variety by divisors, this time using hyperplane sections. This approach is most effective for varieties of small degree. The basic tool here is Siegel's lemma, which will be discussed shortly. In order to illustrate its power we will use it to establish a pair of estimates for the counting functions associated to non-singular del Pezzo surfaces. In Section 7.2 we will prove the following upper bound due to Heath-Brown [74, Theorem 10], adhering to his original argument.

Theorem 7.1. *Let $\varepsilon > 0$ and let $X \subset \mathbb{P}^3$ be a non-singular cubic surface, defined over $\mathbb{Q}$. Let $U \subset X$ be the open subset formed by deleting the 27 lines. Then we have*

$$N_U(B) = O_\varepsilon(B^{\frac{52}{27}+\varepsilon}).$$

It would not be hard, as in the proof of [74, Theorem 10], to extend Theorem 7.1 to the setting of non-singular surfaces in $\mathbb{P}^3$ of arbitrary degree $d \geqslant 3$. This would lead to an exponent $\frac{4}{3}+\frac{16}{9d}+\varepsilon$, together with an additional dependence on d in the implied constant. One notes that this exponent improves (7.1) for smaller values of d, the latter taking over for $d \geqslant 6$.

Next we will focus our attention on non-singular surfaces in $\mathbb{P}^4$ arising as the intersection of two quadrics. The aim in Section 7.3 is to establish Theorem 2.3. Let $S \subset \mathbb{P}^4$ be a non-singular del Pezzo surface of degree 4 and let $U \subset S$ be the open subset obtained by deleting the 16 lines from S. Then we will show that

$$N_U(B) = O_{\varepsilon,S}(B^{\frac{5}{4}+\varepsilon}), \tag{7.2}$$

under the hypothesis that Conjecture 2.7 holds. One notes that Theorem 7.1 is uniform in the coefficients of the defining polynomial, but that (7.2) involves an arbitrary dependence on the surface.

The following result is a simple application of the pigeon-hole principle and lies at the heart of this chapter's discussion.

Lemma 7.1. *Let $n \geqslant 2$ and let $B_1, \dots, B_n > 0$. Let $\mathbf{x} \in \mathbb{Z}^n$ lie in the box $|x_i| \leqslant B_i$, for $1 \leqslant i \leqslant n$. Then there exists a vector $\mathbf{y} \in \mathbb{Z}^n_{\mathrm{prim}}$, for which $\mathbf{x}.\mathbf{y} = 0$ and*

$$|y_i| \leqslant B_i^{-1}(nB_1 \cdots B_n)^{\frac{1}{n-1}} \quad (1 \leqslant i \leqslant n).$$

Proof. Let $\mathscr{B} = (nB_1 \cdots B_n)^{\frac{1}{n-1}}$ and let $\mathbf{x} \in \mathbb{Z}^n$ be such that $|x_i| \leqslant B_i$. Then for any $\mathbf{y}$ satisfying $0 \leqslant y_i \leqslant B_i^{-1}\mathscr{B}$, the modulus of $\mathbf{x}.\mathbf{y}$ is at most $\sum_{i=1}^n B_i[B_i^{-1}\mathscr{B}] \leqslant n\mathscr{B}$. Moreover the total number of available $\mathbf{y}$ is

$$\prod_{i=1}^n([B_i^{-1}\mathscr{B}]+1) > \prod_{i=1}^n B_i^{-1}\mathscr{B} = n\mathscr{B}.$$

Consequently there exist distinct such integer vectors $\mathbf{y}^{(1)}, \mathbf{y}^{(2)}$ such that $\mathbf{x}.\mathbf{y}^{(1)} = \mathbf{x}.\mathbf{y}^{(2)}$. We therefore complete the proof of the lemma by setting $\mathbf{y} = \mathbf{y}^{(1)} - \mathbf{y}^{(2)}$ and removing common factors. □

Taking $B_1 = \cdots = B_n = B$ in Lemma 7.1, we deduce the following consequence.

Lemma 7.2 (Siegel's lemma). *Let $n \geqslant 2$. Then for any $x \in \mathbb{P}^{n-1}(\mathbb{Q})$ such that $H(x) \leqslant B$, there exists a hyperplane $\Lambda = \{\mathbf{a}.\mathbf{x} = 0\} \in \mathbb{P}^{n-1*}(\mathbb{Q})$ of height $O_n(B^{\frac{1}{n-1}})$ such that $x \in \Lambda$.*

An alternative proof of Siegel's lemma can be obtained using the geometry of numbers, as in Exercise 7.1. Even in the setting $n = 3$ this result can be combined with Bézout's theorem to give a non-trivial bound for $N_C(B)$, when $C \subset \mathbb{P}^2$ is an irreducible plane curve of degree $d \geqslant 2$. Thus it follows from Exercise 7.2 that

$$N_C(B) = O_d(B^{\frac{3}{2}}),$$

for such curves C. This improves on the "trivial bound" in Theorem 3.1, but compares rather badly to Theorem 4.2.

7.1 Dual variety

Our work will require some background information on the so-called *dual variety*. Let $X \subset \mathbb{P}^{n-1}$ be a non-singular variety of degree $d \geqslant 3$. Then the dual variety X^* is defined to be the closure of the set of hyperplanes $H \in \mathbb{P}^{n-1^*}$ such that the intersection $H \cap X$ is singular. In fact when X is a hypersurface, X^* is actually equal to the locus of such hyperplanes.

Suppose for the moment that X is a hypersurface, with underlying form $F \in \mathbb{Z}[x_1, \ldots, x_n]$ of degree d. The hypersurface $X_{\mathbf{y}}$, say, obtained from the linear slice

$$F(\mathbf{x}) = 0, \quad \mathbf{y}.\mathbf{x} = 0,$$

is singular if and only if there exists $[\mathbf{z}] \in X$ such that $\mathbf{y}$ is proportional to $\nabla F(\mathbf{z})$, by the Jacobian criterion. This is equivalent to the existence of $\lambda \in \overline{\mathbb{Q}}^*$ such that $\nabla F(\mathbf{z}) = \lambda \mathbf{y}$. Note that we cannot have $\lambda = 0$ since F is assumed to be non-singular. Replacing λ by μ^{d-1}, we may therefore conclude that $X_{\mathbf{y}}$ is singular if and only if there exists a non-zero vector $\mathbf{z} \in \overline{\mathbb{Q}}^n$ such that

$$\mu^{d-1}\mathbf{y} = \nabla F(\mathbf{z}), \quad F(\mathbf{z}) = 0.$$

This defines a system of $n+1$ homogeneous equations in $n+1$ unknowns. The dual variety X^* is obtained via elimination theory as a hypersurface $F^*(\mathbf{y}) = 0$, with $F^* \in \mathbb{Z}[y_1, \ldots, y_n]$ a form of degree $d(d-1)^{n-2}$. It might be instructive to consider a concrete example. Consider the cubic curve $x_1^3 + x_2^3 + x_3^3 = 0$. Then one can check that the dual variety is given by the equation

$$y_1^6 + y_2^6 + y_3^6 - 2y_1^3y_2^3 - 2y_1^3y_3^3 - 2y_2^3y_3^3 = 0.$$

This illustrates the fact that even for non-singular hypersurfaces the dual variety is often singular.

For general non-singular varieties the dual variety need not be a hypersurface, though this is the case for non-singular hypersurfaces and for non-singular varieties arising as the complete intersection of two hypersurfaces of degrees $d_1, d_2 \geqslant 2$. Moreover, it is part of the general theory of the dual variety that it is irreducible and defined over $\mathbb{Q}$ for both of these examples. This is established in [52, Proposition 5.7.2], for example.

Henceforth let X be one of the two types of varieties discussed in the preceding paragraph. Let us write d^* for the degree of X^*. Arguing as in the case of hypersurfaces it is easy to see that $d^* = O_d(1)$. It is also straightforward to deduce that $d^* > 1$. For this we treat only the case of hypersurfaces, the remaining case being a trivial extension. If we suppose that X^* is defined by the equation $\mathbf{a}.\mathbf{y} = 0$ for some $\mathbf{a}$, then

$$\mathbf{a}.\nabla F(\mathbf{x}) = 0$$

for every $[\mathbf{x}] \in X$. Since $d \geqslant 3$ this implies that $\mathbf{a}.\nabla F$ vanishes identically, whence Euler's identities yield

$$dF(\mathbf{a}) = \mathbf{a}.\nabla F(\mathbf{a}) = 0, \quad d\frac{\partial F}{\partial x_i}(\mathbf{a}) = \Big(\frac{\partial}{\partial x_i}(\mathbf{a}.\nabla F)\Big)(\mathbf{a}) = 0,$$

for $1 \leqslant i \leqslant n$. This shows that $\mathbf{a}$ is a singular point of X which is impossible.

The dual variety therefore affords us quite good control over those hyperplane sections of a non-singular variety which produce singular divisors. We collect together our investigation so far in the following result.

Lemma 7.3. *Suppose that $X \subset \mathbb{P}^{n-1}$ is a non-singular variety of degree $d \geqslant 3$, which is either a hypersurface or the intersection of two hypersurfaces of degrees $d_1, d_2 \geqslant 2$. Then the dual variety $X^* \subset \mathbb{P}^{n-1}$ is an irreducible hypersurface of degree d^*, with $2 \leqslant d^* \ll_d 1$, which is defined over $\mathbb{Q}$.*

The theory of dual varieties is discussed more extensively in the work of Zak [126]. In particular if $X \subset \mathbb{P}^{n-1}$ is a non-degenerate non-singular surface of degree d, then it is recorded in [126, Proposition 3.2] that the degree of the dual variety satisfies $d^* \geqslant \max\{d-1, n-3\}$.

7.2 Non-singular del Pezzo surfaces of degree 3

In this section we establish Theorem 7.1. Siegel's lemma shows that any point in $\mathbb{P}^3(\mathbb{Q})$ of height at most B must lie on one of at most $O(B^{\frac{4}{3}})$ planes $\mathbf{x}.\mathbf{y} = 0$, with $\mathbf{y} \in \mathbb{Z}^4$ running over primitive vectors such that $|\mathbf{y}| = O(B^{\frac{1}{3}})$. Suppose that $X \subset \mathbb{P}^3$ is a cubic surface with underlying equation $F(\mathbf{x}) = 0$, with $F \in \mathbb{Z}[x_1, \ldots, x_4]$. The hyperplane sections

$$C_{\mathbf{y}}: \quad \mathbf{x}.\mathbf{y} = F(\mathbf{x}) = 0$$

all produce cubic curves contained in X.

We are interested in points lying on the open subset $U \subset X$ formed by deleting the lines from X. In particular we are not interested in points lying on any linear components of $C_{\mathbf{y}}$ since these will correspond to points lying on one of the lines. For generic $\mathbf{y}$ the curve $C_{\mathbf{y}}$ will be an irreducible cubic curve and so will have $O_\varepsilon(B^{\frac{2}{3}+\varepsilon})$ points to contribute to $N_U(B)$, by Theorem 4.2. This would therefore give an overall contribution of $O_\varepsilon(B^{\frac{4}{3}+\frac{2}{3}+\varepsilon})$ from the generic slices, which is not good enough.

Instead we employ the geometry of numbers to get the desired saving of $\frac{2}{27}$ that is visible in Theorem 7.1. Now Lemma 4.4 implies that for fixed primitive $\mathbf{y}$ the equation $\mathbf{x}.\mathbf{y} = 0$ defines a lattice of rank 3 and determinant $|\mathbf{y}|$. Choose a basis $\mathbf{b}^{(1)}, \mathbf{b}^{(2)}, \mathbf{b}^{(3)}$ for the lattice, with the properties listed in Lemma 4.3. In particular $|\mathbf{b}^{(1)}| \ll |\mathbf{b}^{(2)}| \ll |\mathbf{b}^{(3)}|$ and

$$|\mathbf{y}| \ll |\mathbf{b}^{(1)}||\mathbf{b}^{(2)}||\mathbf{b}^{(3)}| \ll |\mathbf{y}|.$$

Thus it follows
$$|\mathbf{b}^{(1)}| \ll |\mathbf{y}|^{\frac{1}{3}} \ll B^{\frac{1}{9}}. \tag{7.3}$$
We now carry out the non-singular change of variables
$$\mathbf{x} = \lambda_1 \mathbf{b}^{(1)} + \lambda_2 \mathbf{b}^{(2)} + \lambda_3 \mathbf{b}^{(3)},$$
with $\boldsymbol{\lambda} = (\lambda_1, \lambda_2, \lambda_3) \in \mathbb{Z}^3$ such that $\lambda_i \ll |\mathbf{b}^{(i)}|^{-1}|\mathbf{x}|$, for $1 \leqslant i \leqslant 3$. Since $|\mathbf{b}^{(1)}| \ll |\mathbf{b}^{(2)}| \ll |\mathbf{b}^{(3)}|$ and $|\mathbf{x}| \leqslant B$, so it follows that
$$|\boldsymbol{\lambda}| \ll \frac{B}{|\mathbf{b}^{(1)}|}.$$
Moreover if $\mathbf{x}$ is primitive, then it is clear that $\boldsymbol{\lambda}$ is also primitive. Substituting into $F(\mathbf{x}) = 0$ we deduce that $C_{\mathbf{y}}$ can be defined via the equation $G(\boldsymbol{\lambda}) = 0$, say, in which G is a cubic form defined over $\mathbb{Z}$, but which is not necessarily absolutely irreducible. If $\mathbf{y}$ is generic, so that $C_{\mathbf{y}}$ is indeed irreducible, then it now follows from Theorem 4.2 that each such curve contributes
$$\ll_\varepsilon \Big(\frac{B}{|\mathbf{b}^{(1)}|}\Big)^{\frac{2}{3}+\varepsilon},$$
for any $\varepsilon > 0$. It is important to notice that the implied constant is independent of the vectors $\mathbf{b}^{(i)}$ and $\mathbf{y}$.

We proceed to sum this bound over all the $\mathbf{y}$ that arise, for which we will need to count how many vectors $\mathbf{y}$ correspond to a given $\mathbf{b}^{(1)}$. But Lemmas 4.4 and 4.5 together imply that there are $\ll B^{\frac{2}{3}} + |\mathbf{b}^{(1)}|^{-1}B \ll |\mathbf{b}^{(1)}|^{-1}B$ vectors $\mathbf{y}$ in the region $|\mathbf{y}| \ll B^{\frac{1}{3}}$ which are orthogonal to $\mathbf{b}^{(1)}$. Recall the bound for $\mathbf{b}^{(1)}$ from (7.3). It therefore follows that the overall contribution, when we sum over the $\mathbf{y}$ for which $C_{\mathbf{y}}$ is irreducible, is
$$\ll_\varepsilon \sum_{|\mathbf{b}^{(1)}| \ll B^{\frac{1}{9}}} \Big(\frac{B}{|\mathbf{b}^{(1)}|}\Big)^{\frac{5}{3}+\varepsilon} \ll_\varepsilon B^{\frac{5}{3}+\varepsilon}(B^{\frac{1}{9}})^{4-\frac{5}{3}} \ll_\varepsilon B^{\frac{52}{27}+\varepsilon}.$$
This is satisfactory for Theorem 7.1.

It remains to deal with non-generic $\mathbf{y}$. The theory of dual varieties from Section 7.1 shows that there is a non-zero form $F^* \in \mathbb{Z}[y_1, \dots, y_4]$ such that $F^*(\mathbf{y}) = 0$ if and only if $C_{\mathbf{y}}$ is singular. Moreover, F^* has degree k, with $2 \leqslant k \ll 1$ and is absolutely irreducible, by Lemma 7.3. Thus $C_{\mathbf{y}}$ can be singular for only $O_\varepsilon(B^{\frac{2}{3}+\varepsilon})$ values of $\mathbf{y}$, by the proof of Conjecture 3.4 in the case $n = 4$. If $C_{\mathbf{y}}$ is singular but irreducible then Theorem 4.2 implies that $N_{C_{\mathbf{y}}}(B) \ll_\varepsilon B^{\frac{2}{3}+\varepsilon}$, which thereby makes the satisfactory overall contribution $O_\varepsilon(B^{\frac{4}{3}+\varepsilon})$. When $C_{\mathbf{y}}$ fails to be irreducible, then we may safely ignore any linear components, since we are not interested in points lying on lines in the surface. The remaining irreducible components are the conics contained in the surface. Theorem 4.2 therefore implies that we get the overall contribution $O_\varepsilon(B^{\frac{5}{3}+\varepsilon})$ from the conics. This too is satisfactory and therefore completes the proof of Theorem 7.1.

Of course, as indicated in Section 2.3.1, one can do much better than Theorem 7.1 if one is prepared to assume the rank hypothesis recorded in Conjecture 2.7. The biggest contribution above came from the generic hyperplanes. But for these the corresponding cubic curves $C_{\mathbf{y}}$ are all non-singular genus 1 cubics. Thus either $C_{\mathbf{y}}(\mathbb{Q})$ is empty or else $C_{\mathbf{y}}$ is elliptic and one can use Theorem 4.3 to show that there are very few rational points to worry about. This approach ultimately leads to the conditional final bound $O_{\varepsilon,X}(B^{\frac{4}{3}+\varepsilon})$ for non-singular cubic surfaces.

7.3 Non-singular del Pezzo surfaces of degree 4

In this section we wish to extend the ideas of the previous section, by fusing them with the treatment of cubic surfaces in [73], in order to tackle non-singular del Pezzo surfaces of degree 4. Our goal is to establish the estimate (7.2), subject to Conjecture 2.7.

Suppose that $S \subset \mathbb{P}^4$ is given by the zero locus of a pair of quadratic forms $Q_1, Q_2 \in \mathbb{Z}[x_1, \dots, x_5]$ such that the Jacobian matrix $(\nabla Q_1, \nabla Q_2)$ has full rank throughout S. Let $U \subset S$ be the open subset formed by deleting all of the lines from S. As in the proof of Theorem 4.1 we may use Lemma 7.1 to cover all of the rational points in $\mathbb{P}^4(\mathbb{Q})$ of height at most B with hyperplanes $\mathbf{x}.\mathbf{y} = 0$, with $\mathbf{y} \in \mathbb{Z}^5_{\text{prim}}$ such that $|\mathbf{y}| \ll B^{\frac{1}{4}}$. It therefore suffices to consider the contribution from the curves

$$C_{\mathbf{y}}: \quad \mathbf{x}.\mathbf{y} = Q_1(\mathbf{x}) = Q_2(\mathbf{x}) = 0,$$

for each vector $\mathbf{y}$. These curves define curves of degree 4 in $\mathbb{P}^4$, but may fail to be non-singular or even irreducible. We may ignore any linear components of the curve, since these will correspond to lines contained in S.

We begin by dealing with the contribution from generic $\mathbf{y}$. Using $\mathbf{x}.\mathbf{y}=0$ to eliminate one of the variables, we see that the curve $C_{\mathbf{y}}$ is a non-singular intersection of two quadrics in $\mathbb{P}^3$, and so has genus 1. Either $C_{\mathbf{y}}(\mathbb{Q})$ is empty, in which case we are happy, or else $C_{\mathbf{y}}$ is elliptic. In the latter case

$$N_{C_{\mathbf{y}}}(B) \ll_{\varepsilon,S} B^{\varepsilon},$$

by Theorem 4.4, under the assumption that Conjecture 2.7 holds. The overall contribution from generic $\mathbf{y}$ is therefore satisfactory from the point of view of (7.2), since there are $O(B^{\frac{5}{4}})$ possible hyperplanes that arise.

Turning to non-generic $\mathbf{y}$, the theory of the dual variety from Section 7.1 shows that there exists a non-zero form $\Delta \in \mathbb{Z}[x_1, \dots, x_5]$ such that $\Delta(\mathbf{y}) = 0$ if and only if $C_{\mathbf{y}}$ is singular. Moreover, Δ has degree k in the range $2 \leqslant k \ll 1$ and is absolutely irreducible, by Lemma 7.3. It therefore follows from the resolution of Conjecture 3.4 in the case $n = 5$ that $C_{\mathbf{y}}$ can be singular for only $O(B^{\frac{3}{4}})$ values of $\mathbf{y}$. When $C_{\mathbf{y}}$ is singular but irreducible it follows from Theorem 4.2 that

$N_{C_{\mathbf{y}}}(B) \ll_\varepsilon B^{\frac{1}{2}+\varepsilon}$ uniformly in the coefficients of $C_{\mathbf{y}}$. Thus we get an overall contribution of $O_{\varepsilon,S}(B^{\frac{5}{4}+\varepsilon})$ from the quartic curves, subject to the hypothesis that Conjecture 2.7 holds.

Any irreducible component of $C_{\mathbf{y}}$ that is not defined over $\mathbb{Q}$ will contribute just $O(1)$ to the counting function. Indeed, the rational points on the curve must also lie on the curve obtained by taking a Galois conjugate of the coefficients defining the underlying polynomials. But this intersection contains $O(1)$ points by Bézout's theorem. Thus it suffices to fix attention on the irreducible components of $C_{\mathbf{y}}$ that are defined over $\mathbb{Q}$.

Suppose that $C_{\mathbf{y}}$ contains a quadratic component defined over $\mathbb{Q}$. Then S contains a conic defined over $\mathbb{Q}$ and we may apply (2.17) to deduce the better bound $N_U(B) \ll_{\varepsilon,S} B^{1+\varepsilon}$ in this case, which is satisfactory.

The remaining case to consider is the possibility that $C_{\mathbf{y}}$ contains a linear component defined over $\mathbb{Q}$. This must correspond to one of the lines on S, which following a non-singular linear change of variables, we may assume to take the shape $x_1 = x_2 = x_3 = 0$. But then the quadratic forms defining S can be written

$$Q_i(x_1, \dots, x_5) = q_i(x_1, x_2, x_3) + x_4\ell_i(x_1, x_2, x_3) + x_5 m_i(x_1, x_2, x_3),$$

for $i = 1, 2$, with each ℓ_i, m_i a linear form and each q_i a quadratic form. All of these forms are defined over $\mathbb{Z}$. We wish to eliminate x_5 from one of the equations. If m_1 or m_2 is identically zero, then the point $[0, 0, 0, 0, 1]$ would produce a singular point on S, which is impossible. Hence neither m_1 nor m_2 is identically zero. We can now eliminate x_5 from the first equation to get the pair of equations

$$\begin{cases} c(x_1, x_2, x_3) + x_4 q(x_1, x_2, x_3) = 0, \\ q_2(x_1, x_2, x_3) + x_4\ell_2(x_1, x_2, x_3) + x_5 m_2(x_1, x_2, x_3) = 0, \end{cases} \tag{7.4}$$

for linear forms ℓ_2, m_2, quadratic forms q, q_2 and a cubic form c, all of which are defined over $\mathbb{Z}$, with c and q non-zero.

The hyperplane sections in which we are currently interested all pass through the line $x_1 = x_2 = x_3 = 0$. They are defined by vectors $\mathbf{y} = (\alpha, \beta, \gamma, 0, 0)$ for relatively coprime $\alpha, \beta, \gamma \in \mathbb{Z}$ such that $\alpha, \beta, \gamma \ll B^{\frac{1}{4}}$. Assume without loss of generality that $|\alpha| \leqslant |\beta| \leqslant |\gamma|$. Any primitive vector satisfying $\mathbf{x}.\mathbf{y} = 0$ takes the shape

$$(\gamma t_1, \gamma t_2, -\alpha t_1 - \beta t_2, t_3, t_4),$$

for $\mathbf{t} \in \mathbb{Z}^4_{\text{prim}}$. It will suffice to count primitive vectors $\mathbf{t}' = (t_1, t_2, t_3)$ satisfying the first equation in (7.4) since there are only $O(1)$ rational points in the intersection of (7.4) with $\mathbf{x}.\mathbf{y} = 0$ that lie above a given choice of $\mathbf{t}'$. We are therefore charged with the task of counting $\mathbf{t}' \in \mathbb{Z}^3_{\text{prim}}$ for which

$$c(\gamma t_1, \gamma t_2, -\alpha t_1 - \beta t_2) + t_3 q(\gamma t_1, \gamma t_2, -\alpha t_1 - \beta t_2) = 0, \tag{7.5}$$

with

$$|t_1|, |t_2| \leqslant \frac{B}{|\gamma|}, \quad |t_3| \leqslant B.$$

The latter inequalities follow from the fact that we are counting vectors $\mathbf{x} \in \mathbb{Z}^4$ such that $|\mathbf{x}| \leqslant B$. Furthermore, we may assume that (7.5) defines an irreducible cubic curve. By viewing (7.5) modulo γ we deduce that

$$(\alpha t_1 + \beta t_2)^2 (t_3 A - A'(\alpha t_1 + \beta t_2)) \equiv 0 \pmod{\gamma},$$

where $A = q(0,0,1)$ and $A' = c(0,0,1)$. Any prime dividing γ must therefore divide $\alpha t_1 + \beta t_2$ or $t_3 A - A'(\alpha t_1 + \beta t_2)$. Let

$$\xi(\gamma) := \prod_{p \mid \gamma} p$$

denote the square-free kernel of γ. It follows that there is a factorisation $\xi(\gamma) = g_1 g_2$ such that

$$\alpha t_1 + \beta t_2 \equiv 0 \pmod{g_1} \tag{7.6}$$

and

$$t_3 A \equiv A'(\alpha t_1 + \beta t_2) \pmod{g_2}.$$

The first of these congruences forces t_1, t_2 to lie in a sublattice $\Lambda \subseteq \mathbb{Z}^2$ of determinant g_1. By Lemma 4.3 there exist basis vectors $\mathbf{b}^{(1)}, \mathbf{b}^{(2)} \in \Lambda$, with

$$|\mathbf{b}^{(1)}| \ll |\mathbf{b}^{(2)}|, \quad g_1 \ll |\mathbf{b}^{(1)}||\mathbf{b}^{(2)}| \ll g_1, \tag{7.7}$$

such that whenever we write $(t_1, t_2) = u_1 \mathbf{b}^{(1)} + u_2 \mathbf{b}^{(2)}$ for integers u_1, u_2, we automatically have

$$u_1 \ll \frac{B}{|\gamma||\mathbf{b}^{(1)}|}, \quad u_2 \ll \frac{B}{|\gamma||\mathbf{b}^{(2)}|}.$$

To handle the second congruence above we merely make a change of variables

$$t_3 A = A'(\alpha t_1 + \beta t_2) + g_2 u_3,$$

where now u_3 is an integer in the range $u_3 \ll_S g_2^{-1} B$.

Making these changes of variables in (7.5) leads us to an equation of the shape

$$C(u_1, u_2) + u_3 Q(u_1, u_2) = 0,$$

for non-zero integral binary forms C, Q of degrees 3 and 2, respectively. This defines a singular irreducible plane curve of degree 3. However, rather than applying Theorem 4.2 directly, we would like a version of this result which counts solutions that lie in lop-sided boxes. Indeed, we have seen that $u_i \ll B_i$ for $1 \leqslant i \leqslant 3$, with

$$B_1 = \frac{B}{|\gamma||\mathbf{b}^{(1)}|}, \quad B_2 = \frac{B}{|\gamma||\mathbf{b}^{(2)}|}, \quad B_3 = \frac{B}{g_2}. \tag{7.8}$$

In particular $1 \leqslant B_2 \ll B_1 \leqslant B_3$.

Let

$$F(x_1, x_2, x_3) := \sum_{\substack{\mathbf{e}\in\mathbb{Z}^3\\ e_1+e_2+e_3=d}} a_{\mathbf{e}} x_1^{e_1} x_2^{e_2} x_3^{e_3} \in \mathbb{Z}[x_1, x_2, x_3]$$

be an absolutely irreducible form of degree d and recall the definition (4.8) of the counting function $N(F; \mathbf{B})$. The following result, which is due to Heath-Brown [74, Theorem 3], provides a useful upper bound for this quantity.

Lemma 7.4. *Let F be as above, let $\varepsilon > 0$ and let $B_1, B_2, B_3 \geqslant 1$. Then we have*

$$N(F; \mathbf{B}) \ll_{d,\varepsilon} T^{-\frac{1}{d^2}} V^{\frac{1}{d}+\varepsilon},$$

where

$$V := B_1 B_2 B_3, \quad T := \max_{\substack{e_1+e_2+e_3=d\\ a_{\mathbf{e}}\neq 0}} B_1^{e_1} B_2^{e_2} B_3^{e_3}.$$

When $B_1 = B_2 = B_3 = B$, one easily retrieves Theorem 4.2 in the special case $n = 3$. We wish to apply Lemma 7.4 to our problem, for which we take B_1, B_2, B_3 as in (7.8). In particular, it is clear that we can take $T = B_2^2 B_3$. Thus, for fixed α, β, γ, we get a contribution of

$$\ll_{\varepsilon,S} \frac{(B_1 B_2 B_3)^{\frac{1}{3}+\varepsilon}}{B_2^{\frac{2}{9}} B_3^{\frac{1}{9}}} \ll_{\varepsilon,S} \frac{B^{\frac{2}{3}+\varepsilon}}{|\gamma|^{\frac{4}{9}} |\mathbf{b}^{(1)}|^{\frac{1}{3}} |\mathbf{b}^{(2)}|^{\frac{1}{9}} g_2^{\frac{2}{9}}} \ll_{\varepsilon,S} \frac{B^{\frac{2}{3}+\varepsilon}}{|\gamma|^{\frac{4}{9}} |\mathbf{b}^{(1)}|^{\frac{2}{9}} g_1^{\frac{1}{9}} g_2^{\frac{2}{9}}}$$

from the individual cubic curves under consideration. We now need to sum this contribution over all $\alpha, \beta, \gamma \ll B^{\frac{1}{4}}$. Note from (7.7) that $|\mathbf{b}^{(1)}| \ll \sqrt{g_1}$ and recall the definition Λ of the lattice of $(t_1, t_2) \in \mathbb{Z}^2$ such that (7.6) holds. It follows that

$$\begin{aligned}
\sum_{\alpha,\beta} \frac{1}{|\mathbf{b}^{(1)}|^{\frac{2}{9}}} &\ll \sum_{\substack{j\geqslant 0\\ P=2^j \ll \sqrt{g_1}}} \frac{1}{P^{\frac{2}{9}}} \sum_{\substack{\mathbf{b}\in\mathbb{Z}^2\\ P<|\mathbf{b}|\leqslant 2P}} \sum_{\substack{\alpha,\beta\ll B^{\frac{1}{4}}\\ g_1\mid(\alpha b_1+\beta b_1)}} 1 \\
&\ll \frac{B^{\frac{1}{2}}}{g_1} \sum_{\substack{j\geqslant 0\\ P=2^j\ll\sqrt{g_1}}} P^{2-\frac{2}{9}} \\
&\ll \frac{B^{\frac{1}{2}}}{g_1^{\frac{1}{9}}},
\end{aligned}$$

by Lemma 4.5.

On recalling that $g_1 g_2 = \xi(\gamma)$, we are therefore led to the overall contribution

$$\ll_{\varepsilon} \sum_{\gamma} \frac{B^{\frac{7}{6}+\varepsilon}}{\xi(\gamma)^{\frac{2}{3}}}.$$

We estimate this sum using Rankin's trick. Thus for any $G \geqslant 1$ and $\theta \in [0,1]$, we find that

$$\begin{aligned}\sum_{\gamma \leqslant G} \xi(\gamma)^{-\theta} &\leqslant \sum_{\gamma \leqslant G} \xi(\gamma)^{-\theta}\Big(\frac{G}{\gamma}\Big)^{1-\theta+\varepsilon}\\ &\leqslant G^{1-\theta+\varepsilon}\sum_{\gamma=1}^{\infty} \xi(\gamma)^{-\theta}\gamma^{\theta-1-\varepsilon}\\ &\leqslant G^{1-\theta+\varepsilon}\prod_p \Big(1+p^{-\theta}p^{\theta-1-\varepsilon}+p^{-\theta}p^{2\theta-2-2\varepsilon}+\cdots\Big)\\ &= G^{1-\theta+\varepsilon}\prod_p \Big(1+p^{-\theta}\frac{p^{\theta-1-\varepsilon}}{1-p^{\theta-1-\varepsilon}}\Big)\\ &\ll_\varepsilon G^{1-\theta+\varepsilon}.\end{aligned}$$

Applying this with $G = cB^{\frac{1}{4}}$ and $\theta = \frac{2}{3}$, for a suitable absolute constant $c > 0$, therefore leads to the overall contribution

$$\ll_{\varepsilon,S} B^{\frac{7}{6}+\frac{1}{12}+2\varepsilon} = B^{\frac{5}{4}+2\varepsilon}.$$

This too is satisfactory and so completes the proof of (7.2), on redefining the choice of $\varepsilon > 0$.

Exercises for Chapter 7

Exercise 7.1. Use Lemmas 4.3 and 4.4 to give an alternative proof of Siegel's lemma.

Exercise 7.2. Let $G \in \mathbb{Z}[x_1,x_2,x_3]$ be an arbitrary non-zero form of degree d that contains no linear factors, and let $B_i > 0$. Use Lemma 7.1 to show that

$$\#\{\mathbf{x} \in \mathbb{Z}^3 : G(\mathbf{x}) = 0,\ |x_i| \leqslant B_i,\ \gcd(x_1,x_2,x_3) = 1\} \ll_d 1 + (B_1B_2B_3)^{\frac{1}{2}},$$

where the implied constant depends at most on d.

Exercise 7.3. Let $F(\mathbf{x}) = x_1^d + x_2^d + x_3^d + x_4^d = 0$ be the Fermat surface of degree d. Show that the dual surface is defined by the form

$$F^*(\mathbf{u}) = \prod_{\omega_2,\omega_3,\omega_4} \big(u_1^{\frac{d}{d-1}} + \omega_2 u_2^{\frac{d}{d-1}} + \omega_3 u_3^{\frac{d}{d-1}} + \omega_4 u_4^{\frac{d}{d-1}}\big),$$

where the product is over all $(d-1)$th roots of unity $\omega_2, \omega_3, \omega_4$.

Exercise 7.4. Let $\varepsilon > 0$ and let $X \subset \mathbb{P}^3$ be a non-singular surface of degree d, defined over $\mathbb{Q}$. Let $U \subset X$ be the open subset formed by deleting the lines from X. Adapt the proof of Theorem 7.1 to show that

$$N_U(B) = O_{d,\varepsilon}(B^{\frac{4}{3}+\frac{16}{9d}+\varepsilon}).$$

Chapter 8

The Hardy–Littlewood circle method

One of the most significant all-purpose tools available in the study of rational points on higher-dimensional algebraic varieties is the Hardy–Littlewood circle method. In this chapter we will illustrate the power of this technique both as a theoretical tool and as a heuristic tool. In Section 8.2 we will establish Birch's Theorem 1.1 in the case $d = 4$ of quartic forms. Here, as in most applications of the circle method, the number of variables needed is rather large compared to the degree. Nonetheless, the circle method can still be used as a purely heuristic tool when the number of variables is smaller. Thus, in Section 8.3, we will provide some evidence for Manin's Conjecture 2.3 in the setting of diagonal cubic surfaces.

We begin by introducing some essential features of the Hardy–Littlewood method. Let $f \in \mathbb{Z}[x_1, \ldots, x_n]$ be a homogeneous polynomial of degree d and let

$$N(f; B) := \#\{\mathbf{x} \in \mathbb{Z}^n : f(\mathbf{x}) = 0, \ |\mathbf{x}| \leqslant B\},$$

for any $B \geqslant 1$. As highlighted in (1.7) it is a simple matter to go between counting all integer solutions and counting only primitive solutions, as long as the exponent of B in the asymptotic formula for $N(f; B)$ exceeds 1. We will therefore concentrate on the counting function $N(f; B)$ in this discussion.

In what follows let $e(z) := e^{2\pi i z}$ and $e_q(z) := e(\frac{z}{q})$, for any $z \in \mathbb{R}$. We will no longer be concerned with estimates that are uniform in the coefficients of the underlying equations. Henceforth we therefore allow our implied constants to depend in any way upon the equations that are under consideration.

The starting point for the Hardy–Littlewood circle method is the generating function

$$S(\alpha) = S_B(\alpha; f) := \sum_{\mathbf{x} \in \mathbb{Z}^n \cap [-B, B]^n} e(\alpha f(\mathbf{x})),$$

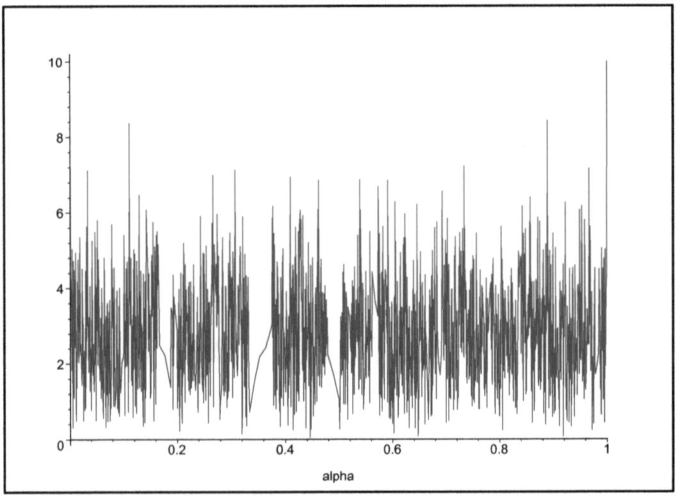

Figure 8.1: Modulus of $S_{10}(\alpha; x^3)$ for $\alpha \in [0,1]$

in which we take $\alpha \in [0,1]$ to be a real variable. The igniting spark is then the simple identity

$$\int_0^1 e(\alpha n)\mathrm{d}\alpha = \begin{cases} 1, & \text{if } n = 0, \\ 0, & \text{if } n \in \mathbb{Z} \setminus \{0\}. \end{cases}$$

Applying this to the generating function $S(\alpha)$ produces the expression

$$N(f; B) = \int_0^1 S(\alpha)\mathrm{d}\alpha. \tag{8.1}$$

The cubic exponential sum $S(\alpha) = S_B(\alpha; f)$ can actually be rather large when α is well-approximated by a rational number $\frac{a}{q}$ with small denominator. This is illustrated in Figure 8.1 for the unary polynomial $f(x) = x^3$. The philosophy that underpins the Hardy–Littlewood method is that one expects $S(\alpha)$ to be small for values of $\alpha \in (0,1)$ that are not well-approximated by rational numbers with small denominator. Showing that this is the case is almost always the most challenging part of any application of the circle method.

Now if $0 < \alpha < 1$, one might expect the numbers $e(\alpha f(\mathbf{x}))$ to be randomly scattered around the unit circle as we range over $\mathbf{x} \in \mathbb{Z}^n$. This leads via the central limit theorem to the expectation that $S(\alpha)$ should usually be of order roughly $B^{\frac{n}{2}}$. Recall from the naive heuristic in (1.3) that the main term in any asymptotic formula should be of order B^{n-d}. Thus we can only hope to succeed with the circle method when $n > 2d$. In most applications we remain rather far from this theoretical limit.

Define the complete exponential sums

$$S_{q,a} := \sum_{\mathbf{x} \pmod{q}} e_q(af(\mathbf{x})). \tag{8.2}$$

Clearly $e_q(af(\mathbf{x})) = e_q(af(\mathbf{y}))$ whenever $\mathbf{x} \equiv \mathbf{y} \pmod{q}$, so that

$$S\Big(\frac{a}{q}\Big) = \sum_{\mathbf{x} \pmod{q}} e_q(af(\mathbf{x}))\#\{\mathbf{y} \in \mathbb{Z}^n : |\mathbf{y}| \leqslant B,\ \mathbf{y} \equiv \mathbf{x} \pmod{q}\}.$$

But Exercise 5.1 reveals that the cardinality in the sum is

$$\Big(\frac{2B}{q} + O(1)\Big)^n = \frac{2^n B^n}{q^n} + O(B^{n-1}),$$

whence

$$S\Big(\frac{a}{q}\Big) = \frac{2^n B^n}{q^n} S_{q,a} + O(q^n B^{n-1}).$$

Therefore, for any fixed fraction $\frac{a}{q}$, we see that $S(\frac{a}{q})$ will be of exact order B^n if $S_{q,a} \neq 0$. It is not hard to extend this analysis, showing that $S(\alpha)$ is also of exact order B^n when α is close to $\frac{a}{q}$. This analysis will form the subject of the following section.

We end this overview of the circle method by noting that everything we have described so far carries over immediately to the more general counting function

$$N(f; B, r, \mathbf{a}, \mathscr{R}) := \#\{\mathbf{x} \in \mathbb{Z}^n : f(\mathbf{x}) = 0,\ \mathbf{x} \in B\mathscr{R},\ \mathbf{x} \equiv \mathbf{a} \pmod{r}\},$$

for fixed $\mathbf{a} \in \mathbb{Z}^n$, any fixed modulus $r \in \mathbb{N}$ such that $\gcd(r, \mathbf{a}) = 1$, and any bounded region $\mathscr{R} \subset \mathbb{R}^n$. Here we write $\mathbf{x} \in B\mathscr{R}$ if and only if $B^{-1}\mathbf{x} \in \mathscr{R}$. To study this more general counting function one simply needs to work with the modified generating function

$$\sum_{\substack{\mathbf{x} \in \mathbb{Z}^n \cap B\mathscr{R} \\ \mathbf{x} \equiv \mathbf{a} \pmod{r}}} e(\alpha f(\mathbf{x})),$$

for $\alpha \in [0, 1]$. The chief advantage of looking at such counting functions lies in their relationship to the problem of determining the distribution of rational points on projective hypersurfaces within the adèles. Studying $N(f; B; r, \mathbf{a}, \mathscr{R})$ directly leads to information on whether or not the hypersurface $f = 0$ satisfies weak approximation.

8.1 Major arcs and minor arcs

Let $a, q \in \mathbb{Z}$ such that $1 \leqslant a \leqslant q \leqslant B$ and $\gcd(a, q) = 1$. Write $\alpha = \frac{a}{q} + z$. Breaking the sum into congruence classes modulo q, as above, we deduce that

$$S\Big(\frac{a}{q} + z\Big) = \sum_{\mathbf{x} \pmod{q}} e_q(af(\mathbf{x})) \sum_{\substack{\mathbf{y} \in \mathbb{Z}^n \cap [-B, B]^n \\ \mathbf{y} \equiv \mathbf{x} \pmod{q}}} e(zf(\mathbf{y})). \tag{8.3}$$

We would like to replace the discrete variable $\mathbf{y}$ by a continuous one in the inner sum, and the summation over $\mathbf{y}$ by an integral. For this we will appeal to the following general result.

Lemma 8.1. *Let $P \geqslant 1$, let $\mathbf{a} \in \mathbb{Z}^n$ and let $r \in \mathbb{N}$ such that $r \leqslant P$. Let F be a function on $\mathbb{R}^n$ all of whose first order partial derivatives exist and are continuous on $\mathscr{R} := [-P, P]^n$. Define*

$$M_F := \sup_{\mathbf{x}\in\mathscr{R}} \max_{1\leqslant i\leqslant n} \Big|\frac{\partial F}{\partial x_i}(\mathbf{x})\Big|.$$

Then we have

$$\sum_{\substack{\mathbf{x}\in\mathbb{Z}^n\cap\mathscr{R}\\ \mathbf{x}\equiv\mathbf{a} \pmod r}} e(F(\mathbf{x})) = \frac{1}{r^n}\int_{\mathscr{R}} e(F(\mathbf{t}))\mathrm{d}\mathbf{t} + O\Big(\frac{P^{n-1}(1+PM_F)}{r^{n-1}}\Big).$$

Proof. Our proof of Lemma 8.1 is based on the Euler–Maclaurin summation formula [117, Section I.0]. Let $B_k(x)$ denote the kth Bernoulli polynomial, for $k \in \mathbb{Z}_{\geqslant 0}$, and let $s \in \mathbb{Z}_{\geqslant 0}$. Let $A, B \in \mathbb{Z}$, with $A < B$. For any function $f : \mathbb{R} \to \mathbb{C}$ whose $(s+1)$-th derivative $f^{(s+1)}$ exists and is continuous on the interval $[A, B]$, the *Euler–Maclaurin summation formula* states that

$$\begin{aligned}\sum_{A<n\leqslant B} f(n) = \int_A^B f(t)\mathrm{d}t + \sum_{k=0}^{s} \frac{(-1)^{k+1}B_{k+1}(0)}{(k+1)!}\big(f^{(k)}(B) - f^{(k)}(A)\big)\\ + \frac{(-1)^s}{(s+1)!}\int_A^B B_{s+1}(t)f^{(s+1)}(t)\mathrm{d}t.\end{aligned} \tag{8.4}$$

Let $a \in \mathbb{Z}$ and $r \in \mathbb{N}$. We will apply this result with $f_0(x) = f(a + rx)$ and

$$A_0 = \frac{A-a}{r}, \quad B_0 = \frac{B-a}{r}.$$

Taking $s = 0$ in the Euler–Maclaurin formula, we therefore deduce that

$$\sum_{\substack{A<n\leqslant B\\ n\equiv a \pmod r}} f(n) = \frac{1}{r}\int_A^B f(t)\mathrm{d}t - \frac{f(B_0)-f(A_0)}{2} + \int_A^B B_1\Big(\frac{t-a}{r}\Big)f'(t)\mathrm{d}t. \tag{8.5}$$

Here, we recall that $B_1(x) = x - [x] - \frac{1}{2}$.

We are now ready to establish Lemma 8.1, which we will do by induction on n. Write $\mathscr{S}_n$ for the n-dimensional sum that is to be estimated. The case $n = 1$ of Lemma 8.1 follows from (8.5) with $f(x) = e(F(x))$. Assuming now that $n \geqslant 2$, we have

$$\mathscr{S}_n = \sum_{\substack{y\in\mathbb{Z}\cap[-P,P]\\ y\equiv a_1 \pmod r}} \sum_{x_2,\dots,x_n} e(G(x_2,\dots,x_n)),$$

where $G(x_2,\dots,x_n) = F(y,x_2,\dots,x_n)$, and the sum over $x_2,\dots,x_n$ is over all integers in $[-P,P]$ such that $x_i \equiv a_i \pmod r$ for $2 \leqslant i \leqslant n$. We may employ the induction hypothesis to estimate the inner sum in $n-1$ variables. It therefore follows that

$$\mathscr{S}_n = \sum_{\substack{y\in\mathbb{Z}\cap[-P,P]\\ y\equiv a_1 \pmod r}} \left(\frac{1}{r^{n-1}} \int_{[-P,P]^{n-1}} e(G(t_2,\dots,t_n))\mathrm{d}t_2\cdots\mathrm{d}t_n + O\Big(\frac{P^{n-2}(1+PM_G)}{r^{n-2}}\Big)\right).$$

Now there are $O(r^{-1}P)$ integers y in the interval $[-P,P]$ that are congruent to a_1 modulo r, since $r \leqslant P$ by assumption. Moreover, it is clear that $M_G \leqslant M_F$. Hence

$$\mathscr{S}_n = \frac{1}{r^{n-1}} \sum_{\substack{y\in\mathbb{Z}\cap[-P,P]\\ y\equiv a_1 \pmod r}} f(y) + O\Big(\frac{P^{n-1}(1+PM_F)}{r^{n-1}}\Big),$$

where

$$f(y) = \int_{[-P,P]^{n-1}} e(F(y,t_2,\dots,t_n))\mathrm{d}t_2\cdots\mathrm{d}t_n.$$

The statement of Lemma 8.1 is now an easy consequence of (8.5). □

Applying Lemma 8.1 in (8.3) we therefore arrive at the following asymptotic formula for $S(\alpha)$, when α is close to a rational number with small denominator.

Lemma 8.2. *Let $a,q \in \mathbb{Z}$ such that $1 \leqslant a \leqslant q \leqslant B$ and $\gcd(a,q)=1$. Then we have*

$$S\Big(\frac{a}{q}+z\Big) = q^{-n}B^n S_{q,a} I(zB^d) + O(qB^{n-1}(1+|z|B^d)),$$

where $S_{q,a}$ is given by (8.2) and

$$I(\gamma) := \int_{\mathbf{t}\in[-1,1]^n} e(\gamma f(\mathbf{t}))\mathrm{d}\mathbf{t}. \tag{8.6}$$

As claimed above, it follows from Lemma 8.2 that for fixed $\frac{a}{q}$ such that $S_{q,a} \neq 0$ the exponential sum $S(\alpha)$ has exact order B^n when $\alpha = \frac{a}{q}+z$ for sufficiently small z. In view of this it makes sense to split the range $[0,1]$ in (8.1) into two subsets, the "major arcs" in which α is close to a rational number with small denominator, and the remaining set, labelled the "minor arcs", on which we would like to show that $S(\alpha)$ is small. This terminology comes from the original formulation of the circle method, in which one integrated around a circle in the complex plane, rather than using the interval $[0,1]$.

There is a good degree of flexibility in the choice of major and minor arcs, but for forms of degree d a natural choice arises through the intervals

$$\mathfrak{M}_{q,a}(\Delta) = \left[\frac{a}{q} - B^{-d+\Delta}, \frac{a}{q} + B^{-d+\Delta}\right]$$

for $1 \leqslant a \leqslant q$ with $\gcd(a, q) = 1$ and $q \leqslant B^{\Delta}$. Here Δ is a small positive parameter which is usually chosen in the final stages of the argument. The full set of major arcs is given by

$$\mathfrak{M}(\Delta) = \bigcup_{1 \leqslant q \leqslant B^{\Delta}} \bigcup_{\substack{1 \leqslant a \leqslant q \\ \gcd(a,q)=1}} \mathfrak{M}_{q,a}(\Delta). \tag{8.7}$$

We take the minor arcs to be defined modulo 1 via $\mathfrak{m}(\Delta) = [0, 1] \setminus \mathfrak{M}(\Delta)$. The following result shows that the intervals in $\mathfrak{M}(\Delta)$ do not overlap if Δ is sufficiently small.

Lemma 8.3. *Let $\Delta < \frac{d}{3}$. Then the intervals $\mathfrak{M}_{q,a}(\Delta)$ are disjoint if $B > 2^{\frac{1}{d-3\Delta}}$.*

Proof. Suppose that $\frac{a}{q} \neq \frac{a'}{q'}$ are distinct centres of two intervals and let $t \in \mathfrak{M}_{q,a}(\Delta) \cap \mathfrak{M}_{q',a'}(\Delta)$. Then

$$\left|\frac{a}{q} - \frac{a'}{q'}\right| \geqslant \frac{1}{qq'} \geqslant \frac{1}{B^{2\Delta}}.$$

However, we also have

$$\left|\frac{a}{q} - \frac{a'}{q'}\right| \leqslant \left|t - \frac{a}{q}\right| + \left|t - \frac{a'}{q'}\right| \leqslant 2B^{-d+\Delta}.$$

These two bounds produce a contradiction if $\Delta < \frac{d}{3}$ and $B > 2^{\frac{1}{d-3\Delta}}$. □

We are now ready to put some of the ideas that we have discussed into practice.

8.2 Quartic hypersurfaces

We now set $d = 4$ in the preceding discussion, taking $f \in \mathbb{Z}[x_1, \ldots, x_n]$ to be a non-singular quartic form. Assume that

$$f(\mathbf{x}) = \sum_{i,j,k,\ell=1}^{n} f_{ijk\ell} x_i x_j x_k x_\ell,$$

for $f_{ijk\ell} \in \mathbb{Z}$ that are symmetric in the indices. We may then define the trilinear forms

$$L_i(\mathbf{w}; \mathbf{x}; \mathbf{y}) := 4! \sum_{j,k,\ell=1}^{n} f_{ijk\ell} w_j x_k z_\ell, \tag{8.8}$$

for $1 \leqslant i \leqslant n$. These will feature quite heavily in the proof. Our goal is to give a reasonably detailed account of the asymptotic formula

$$N(f;B) = c_f B^{n-4}(1+o(1)), \tag{8.9}$$

when $n > 48$, for an appropriate constant c_f to be defined in due course. This is the content of Theorem 1.1 in the case $d = 4$. In recent work of Browning and Heath-Brown [34] this lower bound for n has been improved to $n > 40$ using a more sophisticated treatment of the minor arcs. We will indicate some of the key innovations in Section 8.2.3.

A remarkable feature of the proof of (8.9) is the lack of geometrical input. Almost the only geometric property that is required is recorded in the following result ([6, Lemma 3.3]).

Lemma 8.4. *Define the affine algebraic variety*

$$V = \{(\mathbf{w},\mathbf{x},\mathbf{y}) \in \mathbb{A}^{3n} : \ L_i(\mathbf{w};\mathbf{x};\mathbf{y}) = 0 \ \forall i \leqslant n\}.$$

Then we have $\dim V \leqslant 2n$.

Proof. Consider the diagonal

$$D = \{(\mathbf{w},\mathbf{x},\mathbf{y}) \in \mathbb{A}^{3n} : \ \mathbf{w}=\mathbf{x}=\mathbf{y}\}.$$

Then D has dimension n and it is clear that the intersection $D \cap V$ consists of all points $(\mathbf{x},\mathbf{x},\mathbf{x}) \in \mathbb{A}^{3n}$ for which $\nabla f(\mathbf{x}) = \mathbf{0}$. Thus it follows that $D \cap V$ has affine dimension at most 0 in $\mathbb{A}^{3n}$, since f is non-singular, whence

$$0 \geqslant \dim D\cap V \geqslant \dim D + \dim V - 3n = \dim V - 2n,$$

by the affine dimension theorem. This completes the proof of Lemma 8.4. □

We begin with the identity (8.1). We will break the interval $[0,1]$ into the major arcs and minor arcs described in (8.7) with $d = 4$ and Δ an unspecified constant. From this point onwards any order constants are allowed to depend on Δ, as well as on f. We will assume that $0 < \Delta < 1$ and $B \gg 1$, so that Lemma 8.3 ensures the intervals $\mathfrak{M}_{q,a}(\Delta)$ do not overlap. The volume of $\mathfrak{M}(\Delta)$ is clearly $O(B^{-4+3\Delta})$.

8.2.1 The minor arcs

The central idea in Birch's treatment of the minor arcs is an application of Weyl differencing. Thus one is able to pass from the bound

$$|S(\alpha)|^2 \leqslant \sum_{|\mathbf{w}|\leqslant B} \Big| \sum_{\mathbf{x}\in\mathscr{R}(\mathbf{w})} e\big(\alpha(f(\mathbf{w}+\mathbf{x}) - f(\mathbf{x}))\big)\Big|$$

via Cauchy's inequality to

$$|S(\alpha)|^4 \ll B^n \sum_{|\mathbf{w}|,|\mathbf{x}|\leqslant B} \Big| \sum_{\mathbf{y}\in\mathscr{S}(\mathbf{w},\mathbf{x})} e\big(\alpha f(\mathbf{w},\mathbf{x};\mathbf{y})\big)\Big|, \tag{8.10}$$

where

$$f(\mathbf{w},\mathbf{x};\mathbf{y}) := f(\mathbf{w}+\mathbf{x}+\mathbf{y}) - f(\mathbf{w}+\mathbf{y}) - f(\mathbf{x}+\mathbf{y}) + f(\mathbf{y}).$$

Here $\mathscr{R}(\mathbf{w})$ and $\mathscr{S}(\mathbf{w},\mathbf{x})$ are certain cubes inside $[-B,B]^n$, depending on the arguments. For example $\mathscr{R}(\mathbf{w})$ is just the set of $\mathbf{x}\in[-B,B]^n$ for which also $\mathbf{x}+\mathbf{w}\in[-B,B]^n$. We now repeat this procedure once more on the inner sum in (8.10). This therefore leads to the conclusion that

$$|S(\alpha)|^8 \ll B^{4n} \sum_{|\mathbf{w}|,|\mathbf{x}|,|\mathbf{y}|\leqslant B} \Big| \sum_{\mathbf{z}\in\mathscr{T}(\mathbf{w},\mathbf{x},\mathbf{y})} e\big(\alpha f(\mathbf{w},\mathbf{x},\mathbf{y};\mathbf{z})\big)\Big|, \tag{8.11}$$

where

$$\begin{aligned} f(\mathbf{w},\mathbf{x},\mathbf{y};\mathbf{z}) :=& f(\mathbf{w}+\mathbf{x}+\mathbf{y}+\mathbf{z}) - f(\mathbf{w}+\mathbf{x}+\mathbf{z}) - f(\mathbf{w}+\mathbf{y}+\mathbf{z}) \\ & - f(\mathbf{x}+\mathbf{y}+\mathbf{z}) + f(\mathbf{w}+\mathbf{z}) + f(\mathbf{x}+\mathbf{z}) + f(\mathbf{y}+\mathbf{z}) - f(\mathbf{z}), \end{aligned}$$

and $\mathscr{T}(\mathbf{x},\mathbf{y},\mathbf{z})$ is a certain cube inside $[-B,B]^n$. If we now recall the definition (8.8) of the trilinear forms $L_i(\mathbf{w};\mathbf{x};\mathbf{y})$ for $1\leqslant i\leqslant n$, then we see that

$$f(\mathbf{w},\mathbf{x},\mathbf{y};\mathbf{z}) = \sum_{i=1}^n z_i L_i(\mathbf{w};\mathbf{x};\mathbf{y}) + g(\mathbf{w},\mathbf{x},\mathbf{y}),$$

where $g(\mathbf{w},\mathbf{x},\mathbf{y})$ is independent of $\mathbf{z}$. It therefore follows from (8.11) that

$$\begin{aligned} |S(\alpha)|^8 &\ll B^{4n} \sum_{|\mathbf{w}|,|\mathbf{x}|,|\mathbf{y}|\leqslant B} \Big| \sum_{\mathbf{z}\in\mathscr{T}(\mathbf{w},\mathbf{x},\mathbf{y})} e\Big(\alpha \sum_{i=1}^n z_i L_i(\mathbf{w};\mathbf{x};\mathbf{y})\Big)\Big| \\ &\ll B^{4n} \sum_{|\mathbf{w}|,|\mathbf{x}|,|\mathbf{y}|\leqslant B} \prod_{i=1}^n \min\{B, \|\alpha L_i(\mathbf{w};\mathbf{x};\mathbf{y})\|^{-1}\}, \end{aligned} \tag{8.12}$$

where $\|\beta\|$ denotes the distance to the nearest integer from any $\beta\in\mathbb{R}$. The latter bound follows from the simple estimate

$$\sum_{a<n\leqslant b} e(\beta n) \ll \min\{|b-a|, \|\beta\|^{-1}\},$$

which is proved by viewing the sum as a geometric series.

We proceed to define the quantity

$$N(\alpha,B) := \#\{(\mathbf{w},\mathbf{x},\mathbf{y})\in(\mathbb{Z}\cap[-B,B])^{3n} : \ \|\alpha L_i(\mathbf{w};\mathbf{x};\mathbf{y})\| < B^{-1}\ \forall i\leqslant n\}.$$

It is then a simple matter to deduce that

$$\sum_{|\mathbf{w}|,|\mathbf{x}|,|\mathbf{y}|\leqslant B}\ \prod_{i=1}^{n}\min\{B,\|\alpha L_i(\mathbf{w};\mathbf{x};\mathbf{y})\|^{-1}\} \ll (B\log B)^n N(\alpha,B),$$

as in the proof of [50, Lemma 13.2]. On inserting this into (8.12), it therefore follows that

$$|S(\alpha)|^8 \ll B^{5n}(\log B)^n N(\alpha,B). \tag{8.13}$$

In particular it is clear from the trivial upper bound $N(\alpha,B)\ll B^{3n}$ that we have lost very little in formulating (8.13). In order to handle the quantity $N(\alpha,B)$, appeal is made to the following result of Davenport [50, Lemma 12.6], which is proved using the geometry of numbers.

Lemma 8.5. *Let* $\mathbf{L}$ *be a real symmetric* $n\times n$ *matrix. Let* $A>0$ *be real, and let*

$$N(Z) := \#\{\mathbf{u}\in\mathbb{Z}^n:\ |\mathbf{u}|\leqslant AZ,\ \|(\mathbf{Lu})_i\| < A^{-1}Z\ \forall i\leqslant n\}.$$

Then, if $0<Z_1\leqslant Z_2\leqslant 1$, *we have*

$$N(Z_2)\ll \Big(\frac{Z_2}{Z_1}\Big)^n N(Z_1).$$

The idea is now to apply this result three times, in order to reduce the analysis of $N(\alpha,P)$ to a problem involving the system of equations $L_i(\mathbf{w};\mathbf{x};\mathbf{y})=0$, for $1\leqslant i\leqslant n$. To begin with one takes the matrix $\mathbf{L}$ in Lemma 8.5 to be given by $(\mathbf{Ly})_i = L_i(\mathbf{w};\mathbf{x};\mathbf{y})$. Choosing $A=B$, we have

$$N(1)=\#\{\mathbf{y}\in\mathbb{Z}^n: |\mathbf{y}|\leqslant B,\ \|\alpha L_i(\mathbf{w};\mathbf{x};\mathbf{y})\| < B^{-1}\ \forall i\leqslant n\},$$

and the lemma implies that $N(1)\ll Z^{-n}N(Z)$ for any $0<Z\leqslant 1$. It follows that

$$N(\alpha,B)\ll Z^{-n}\#\Big\{(\mathbf{w},\mathbf{x},\mathbf{y})\in\mathbb{Z}^{3n}:\ \begin{array}{l}|\mathbf{w}|,|\mathbf{x}|\leqslant B,\ |\mathbf{y}|\leqslant ZB,\\ \|\alpha L_i(\mathbf{w};\mathbf{x};\mathbf{y})\| < ZB^{-1}\ \forall i\leqslant n\end{array}\Big\}.$$

Rather than using this estimate directly, Birch permutes the rôles of $\mathbf{w},\mathbf{x}$ and $\mathbf{y}$ in the above. Taking $\mathbf{L}$ to be the matrix given by $(\mathbf{Lx})_i = L_i(\mathbf{w};\mathbf{x};\mathbf{y})$, one applies Lemma 8.5 with $A=BZ^{-\frac12}$, $Z_2=Z^{\frac12}$ and $Z_1=Z^{\frac32}$, in order to deduce that

$$N(\alpha,B)\ll Z^{-2n}\#\Big\{(\mathbf{w},\mathbf{x},\mathbf{y})\in\mathbb{Z}^{3n}:\ \begin{array}{l}|\mathbf{w}|\leqslant B,\ |\mathbf{x}|,|\mathbf{y}|\leqslant ZB,\\ \|\alpha L_i(\mathbf{w};\mathbf{x};\mathbf{y})\| < Z^2B^{-1}\ \forall i\leqslant n\end{array}\Big\}.$$

Finally, it remains to use Lemma 8.5 to shrink the size of the box that $\mathbf{w}$ lies in. Taking $\mathbf{L}$ to be the matrix given by $(\mathbf{Lw})_i = L_i(\mathbf{w};\mathbf{x};\mathbf{y})$, therefore, we apply Lemma 8.5 with $A=BZ^{-1}$, $Z_2=Z$ and $Z_1=Z^2$, to conclude that

$$N(\alpha,B)\ll Z^{-3n}\#S(ZB,Z^{-3}B).$$

Here we have set

$$S(R,Q) := \{(\mathbf{w},\mathbf{x},\mathbf{y}) \in (\mathbb{Z}\cap[-R,R])^{3n} : \|\alpha L_i(\mathbf{w};\mathbf{x};\mathbf{y})\| < Q^{-1}\ \forall i \leqslant n\},$$

for any $R,Q>0$.

The idea now is to find conditions on α and Z under which

$$L_i(\mathbf{w};\mathbf{x};\mathbf{y}) = 0, \quad (1 \leqslant i \leqslant n), \tag{8.14}$$

for every $(\mathbf{w},\mathbf{x},\mathbf{y}) \in S(ZB, Z^{-3}B)$. To do so requires the following result due to Heath-Brown [78, Lemma 2.3], whose proof is so short that we include it here.

Lemma 8.6. *Let $M>0$ and let $\alpha = \frac{a}{q}+z$, with $|z| \leqslant (2qM)^{-1}$. Suppose that $m\in\mathbb{Z}$ is such that $|m|\leqslant M$ and $\|\alpha m\| < Q^{-1}$ for some $Q \geqslant 2q$. Then $m=0$ if in addition we have either $M<q$ or $|z|>(qQ)^{-1}$.*

Proof. To prove the lemma we repeatedly use the fact that $\|x+y\| \leqslant \|x\|+\|y\|$ for every real $x,y\in\mathbb{R}$. If $\|\alpha m\| < Q^{-1}$, then

$$\|\frac{a}{q}m\| \leqslant \|\alpha m\| + \|zm\| < Q^{-1} + (2qM)^{-1}|m| \leqslant (2q)^{-1} + (2qM)^{-1}M = \frac{1}{q}.$$

It follows that $q \mid m$. In particular, if $|m| \leqslant M < q$, then we must have $m=0$. Alternatively, if $|z| > (qQ)^{-1}$, then we observe that

$$|zm| \leqslant (2qM)^{-1}|m| \leqslant (2qM)^{-1}M \leqslant \frac{1}{2},$$

whence

$$|zm| = \|zm\| \leqslant \|\frac{a}{q}m\| + \|\alpha m\| = \|\alpha m\| \leqslant Q^{-1},$$

on recalling that $q \mid m$. Thus $|m| \leqslant (Q|z|)^{-1} < q$, and so $m=0$. This completes the proof of the lemma. ☐

We will use Lemma 8.6 to reduce our consideration to the system of trilinear equations (8.14). Write $\varphi = 24\sum|f_{ijk\ell}|$, where $f_{ijk\ell}$ are the coefficients of f, and suppose that $\alpha = \frac{a}{q}+z$. Then on choosing Z to satisfy the conditions

$$0 < Z \leqslant 1, \quad Z^3 \leqslant (2\varphi q|z|B^3)^{-1}, \quad Z^3 \leqslant \frac{B}{2q} \tag{8.15}$$

and

$$Z^3 < \max\Big\{\frac{q}{\varphi B^3}, q|z|B\Big\}, \tag{8.16}$$

we may make Lemma 8.6 applicable. It follows that

$$N(\alpha,B) \ll Z^{-3n}\#T(ZB),$$

where

$$T(R) := \{(\mathbf{w}, \mathbf{x}, \mathbf{y}) \in (\mathbb{Z} \cap [-R, R])^{3n} : L_i(\mathbf{w}; \mathbf{x}; \mathbf{y}) = 0 \ \forall i \leqslant n\}$$

for any $R > 0$. We are now led to study the density of integer solutions to the system of equations (8.14). Applying Lemma 8.4 we see that this system has affine dimension at most $2n$. Thus (3.1) implies that $\#T(ZB) \ll (ZB)^{2n}$, provided that $ZB \geqslant 1$. This gives

$$N(\alpha, B) \ll Z^{-n} B^{2n}, \tag{8.17}$$

which clearly holds trivially if $ZB < 1$.

We will need to choose Z as large as possible, given the constraints in (8.15) and (8.16). The choice

$$Z = \frac{1}{2} \min \Big\{1, \frac{1}{2\varphi q|z|B^3}, \frac{B}{2q}, \max\Big\{\frac{q}{\varphi B^3}, q|z|B\Big\}\Big\}^{\frac{1}{3}}$$

is clearly satisfactory. On taking this value in (8.13) and (8.17), we therefore deduce that

$$|S(\alpha)|^8 \ll_\varepsilon B^{8n+\varepsilon} \Big(B^{-3} + q|z| + qB^{-4} + q^{-1} \min\Big\{1, \frac{1}{|z|B^4}\Big\}\Big)^{\frac{n}{3}},$$

for any $\varepsilon > 0$, whence

$$S(\alpha) \ll_\varepsilon B^{n+\varepsilon} \Big(B^{-3} + q|z| \max\{1, \frac{1}{|z|B^4}\} + q^{-1} \min\Big\{1, \frac{1}{|z|B^4}\Big\}\Big)^{\frac{n}{24}}. \tag{8.18}$$

We will derive three basic estimates from this bound. The first involves the complete exponential sum $S_{q,a}$ defined in (8.2). This arises by taking $z = 0, B = q$ in the definition of $S(\alpha)$ and repeating the argument above with the box $[-B, B]^n$ replaced by $(0, q]^n$.

An application of (8.18) immediately gives

$$S_{q,a} \ll_\varepsilon q^{\frac{23n}{24}+\varepsilon}, \tag{8.19}$$

for any coprime integers a, q such that $1 \leqslant a \leqslant q$. Next, we claim that

$$S(\alpha) \ll_\varepsilon B^{n+\varepsilon}(|\alpha|B^4)^{-\frac{n}{24}}, \quad \text{if } |\alpha| < B^{-2}. \tag{8.20}$$

This is trivial if $|\alpha| \leqslant B^{-4}$. If $|\alpha| > B^{-4}$, then it follows from (8.18) with $a = 0$, $q = 1$ and $\alpha = z$. Finally, it is a simple matter to deduce the following result from (8.18).

Lemma 8.7. *Let $\varepsilon > 0$ and let a, q, z be such that*

$$1 \leqslant a \leqslant q \leqslant B^2, \quad \gcd(a, q) = 1, \quad |z| \leqslant \frac{1}{q^2}.$$

Then we have

$$S\Big(\frac{a}{q} + z\Big) \ll_\varepsilon B^{n+\varepsilon} q^{-\frac{n}{24}} \min\{1, |z|B^4\}^{-\frac{n}{24}}.$$

We are now ready to conclude the treatment of the minor arcs, producing our analogue of [6, Lemma 4.4]. Let $\alpha \in \mathfrak{m}(\Delta)$. By Dirichlet's approximation theorem we can find coprime integers a, q such that $1 \leqslant a \leqslant q \leqslant B^2$ such that $|q\alpha - a| \leqslant B^{-2}$. In order to be contained in $\mathfrak{m}(\Delta)$ we may assume furthermore that $q > B^{\Delta}$ or $|q\alpha - a| > qB^{-4+\Delta}$. Writing $\alpha = \frac{a}{q} + z$ it therefore follows from Lemma 8.7 that

$$\begin{aligned}\int_{\mathfrak{m}(\Delta)} |S(\alpha)|\mathrm{d}\alpha &\ll_\varepsilon B^{n+\varepsilon} \sum_{q \leqslant B^2} q^{1-\frac{n}{24}} \int_{\frac{-1}{qB^2}}^{\frac{1}{qB^2}} \min\{1, |z|B^4\}^{-\frac{n}{24}} \mathrm{d}z \\ &\ll_\varepsilon B^{n-4-\Delta(\frac{n}{24}-2)+\varepsilon},\end{aligned}$$

under the assumption that $n > 48$. Choosing ε to be sufficiently small in terms of Δ, this therefore shows that the minor arc contribution is $o(B^{n-4})$, as $B \to \infty$. This is satisfactory for (8.9).

8.2.2 The major arcs

It remains to discuss Birch's treatment of the major arcs. Recall the definition (8.2) of the complete exponential sums $S_{q,a}$, for given coprime a, q such that $1 \leqslant a \leqslant q$. Let

$$\mathfrak{S}(R) := \sum_{q \leqslant R} \frac{1}{q^n} \sum_{\substack{a=1 \\ \gcd(a,q)=1}}^{q} S_{q,a}, \tag{8.21}$$

for any $R > 1$. Then we define $\mathfrak{S} = \lim_{R\to\infty} \mathfrak{S}(R)$, when it exists. This is the so-called *singular series* for the problem and encodes information about whether or not the quartic hypersurface has p-adic solutions for all primes p. The real solubility is part and parcel of the so-called *singular integral*, which is here defined to be $\mathfrak{J} := \lim_{R\to\infty} \mathfrak{J}(R)$, when it exists, where

$$\mathfrak{J}(R) := \int_{-R}^{R} \int_{[-1,1]^n} e(zf(\mathbf{x}))\mathrm{d}\mathbf{x}\mathrm{d}z. \tag{8.22}$$

Beginning with the absolute convergence of the singular series, it follows from (8.19) that $\mathfrak{S}$ is absolutely convergent for $n > 48$, and furthermore,

$$\mathfrak{S}(R) = \mathfrak{S} + O_\varepsilon(R^{2-\frac{n}{24}+\varepsilon}), \tag{8.23}$$

for any $\varepsilon > 0$. Recall the definition (8.6) of $I(\gamma)$. The convergence of the singular integral will be a consequence of the following result.

Lemma 8.8. *We have $I(\gamma) \ll_\varepsilon \min\{1, |\gamma|^{-\frac{n}{24}+\varepsilon}\}$ for any $\varepsilon > 0$.*

Proof. The estimate $I(\gamma) \ll 1$ is trivial. In proving the second estimate we may clearly assume that $|\gamma| > 1$. Now it follows from Lemma 8.2 that

$$S(\alpha) = q^{-n}B^n S_{q,a} I(zB^4) + O(|z|qB^{n+3} + qB^{n-1}). \tag{8.24}$$

Taking $a = 0$ and $q = 1$ we therefore deduce that

$$S(\alpha) = B^n I(\alpha B^4) + O\big((|\alpha|B^4 + 1)B^{n-1}\big),$$

for any $B \geqslant 1$. On the other hand, assuming that $|\alpha| < B^{-2}$, (8.20) gives

$$S(\alpha) \ll_\varepsilon B^{n+\varepsilon}(|\alpha|B^4)^{-\frac{n}{24}}.$$

Writing $\alpha B^4 = \gamma$, we may combine these estimates to obtain

$$I(\gamma) \ll_\varepsilon |\gamma|^{-\frac{n}{24}} B^\varepsilon + |\gamma| B^{-1},$$

when $|\gamma| < B^2$. Finally we observe that $I(\gamma)$ is independent of B. Thus we are free to choose $B = |\gamma|^{1+\frac{n}{24}}$, from which Lemma 8.8 follows. $\square$

It now follows from Lemma 8.8 that

$$\mathfrak{I} - \mathfrak{I}(R) = \int_{|\gamma|\geqslant R} I(\gamma)\mathrm{d}\gamma \ll_\varepsilon \int_R^\infty \min\{1, \gamma^{-\frac{n}{24}+\varepsilon}\}\mathrm{d}\gamma \ll_\varepsilon R^{1-\frac{n}{24}+\varepsilon}. \tag{8.25}$$

This shows in particular that $\mathfrak{I}$ is absolutely convergent for $n > 48$.

We are now ready to complete our analysis of the exponential sum $S(\alpha)$ on the major arcs. Let $\alpha = \frac{a}{q} + z \in \mathfrak{M}(\Delta)$, in the notation of (8.7) with $d = 4$. Then a return to (8.24) reveals that

$$S(\alpha) = q^{-n} B^n S_{q,a} I(zB^4) + O(B^{n-1+2\Delta}),$$

since $|z| \leqslant B^{-4+\Delta}$ and $q \leqslant B^\Delta$ on the major arcs. Since the major arcs have measure $O(B^{-4+3\Delta})$, it is now a trivial matter to deduce that

$$\int_{\mathfrak{M}(\Delta)} S(\alpha)\mathrm{d}\alpha = B^{n-4}\mathfrak{S}(B^\Delta)\mathfrak{I}(B^\Delta) + O(B^{n-5+5\Delta}),$$

where $\mathfrak{S}(B^\Delta), \mathfrak{I}(B^\Delta)$ are given by (8.21) and (8.22), respectively.

Note from (8.25) that $\mathfrak{I}(B^\Delta) \ll 1$. Applying (8.23) and (8.25) we therefore deduce that

$$\begin{aligned}\int_{\mathfrak{M}(\Delta)} S(\alpha)\mathrm{d}\alpha &= \mathfrak{S}B^{n-4}\mathfrak{I}(B^\Delta) + O\big(B^{n-5+5\Delta} + B^{n-4-\Delta(\frac{n}{24}-2)}\big)\\ &= \mathfrak{S}\mathfrak{I}B^{n-4} + O_\varepsilon\big(B^{n-5+5\Delta} + B^{n-4-\frac{\Delta}{24}+\varepsilon}\big),\end{aligned}$$

assuming that $n > 48$. Taking Δ such that $0 < \Delta < \frac{1}{5}$ we deduce that there is a contribution of

$$\int_{\mathfrak{M}(\Delta)} S(\alpha)\mathrm{d}\alpha = \mathfrak{S}\mathfrak{I}B^{n-4} + o(B^{n-4})$$

from the major arcs. Once coupled with the treatment of the minor arcs this is enough to establish the asymptotic formula in (8.9) with $c_f = \mathfrak{S}\mathfrak{I}$.

Of course, from the point of view of existence of rational points on the quartic hypersurfaces, one is interested in conditions sufficient to ensure that the constant c_f is positive. We will not discuss the methods that are required for this here, although a closer examination of the singular series is pivotal in our work on diagonal cubic forms in the following section. It turns out that in order to have $c_f > 0$ it is sufficient that the equation $f = 0$ has solutions over $\mathbb{R}$ and over every p-adic field $\mathbb{Q}_p$. This part of the argument is fairly standard and the reader may wish to consult [6] for the details. Assuming this to be the case it therefore follows that the class of non-singular quartic hypersurfaces over $\mathbb{Q}$, in $n > 48$ variables, satisfies the Hasse principle.

8.2.3 Improving Birch's argument

The main bottleneck in improving Birch's argument comes in the treatment of the minor arcs. Indeed, in Exercise 8.1, one shows that the major arcs can be handled for roughly half as many variables. The approach discussed in Section 8.2.1 is based on three successive applications of Weyl differencing, in order to relate the size of the exponential sum $S(\alpha)$ to the locus of integral points on the affine variety cut out by the system of equations $L_i(\mathbf{w};\mathbf{x};\mathbf{y}) = 0$, for $1 \leqslant i \leqslant n$. This approach is quite wasteful, a fact that is capitalised on in the work of Browning and Heath-Brown [34], where the lower bound $n > 48$ is reduced to $n > 40$.

The key idea is to use a differencing argument only once, based instead on the van der Corput method, in order to relate the size of $S(\alpha)$ to the size of a certain family of cubic exponential sums. Let $\alpha \in \mathbb{R}$ and let $H \in [1, B] \cap \mathbb{Z}$. Write, temporarily,

$$\varphi(\mathbf{x}) = \begin{cases} e(\alpha f(\mathbf{x})), & \text{if } |\mathbf{x}| \leqslant B, \\ 0, & \text{otherwise.} \end{cases}$$

The kernel of the van der Corput method is the observation that

$$\#\mathscr{H} S(\alpha) = \sum_{\mathbf{h}\in\mathscr{H}} \sum_{\mathbf{x}\in\mathbb{Z}^n} \varphi(\mathbf{x}+\mathbf{h}) = \sum_{\mathbf{x}\in\mathbb{Z}^n} \sum_{\mathbf{h}\in\mathscr{H}} \varphi(\mathbf{x}+\mathbf{h}),$$

where $\mathscr{H}$ is the set of $\mathbf{h} \in \mathbb{N}^n$ such that $0 < h_i \leqslant H$ for $1 \leqslant i \leqslant n$. An application of Cauchy's inequality yields

$$H^{2n}|S(\alpha)|^2 \ll B^n \sum_{\mathbf{x}\in\mathbb{Z}^n} \Big| \sum_{\mathbf{h}\in\mathscr{H}} \varphi(\mathbf{x}+\mathbf{h}) \Big|^2 = B^n \sum_{\mathbf{h}_1,\mathbf{h}_2\in\mathscr{H}} \sum_{\mathbf{x}\in\mathbb{Z}^n} \varphi(\mathbf{x}+\mathbf{h}_1)\overline{\varphi(\mathbf{x}+\mathbf{h}_2)}.$$

Making the substitution $\mathbf{y} = \mathbf{x} + \mathbf{h}_2$ we deduce that

$$\begin{aligned} H^{2n}|S(\alpha)|^2 &\ll B^n \sum_{\mathbf{h}_1\in\mathscr{H}} \sum_{\mathbf{h}_2\in\mathscr{H}} \sum_{\mathbf{y}\in\mathbb{Z}^n} \varphi(\mathbf{y}+\mathbf{h}_1-\mathbf{h}_2)\overline{\varphi(\mathbf{y})} \\ &= B^n \sum_{\substack{\mathbf{h}\in\mathbb{Z}^n \\ |\mathbf{h}|\leqslant H}} N(\mathbf{h}) \sum_{\mathbf{y}\in\mathbb{Z}^n} \varphi(\mathbf{y}+\mathbf{h})\overline{\varphi(\mathbf{y})}, \end{aligned}$$

where $N(\mathbf{h}) := \#\{\mathbf{h}_1, \mathbf{h}_2 \in \mathscr{H} : \mathbf{h} = \mathbf{h}_1 - \mathbf{h}_2\} \ll H^n$. We therefore conclude that

$$|S(\alpha)|^2 \ll H^{-n} B^n \sum_{\mathbf{h}} |T_{\mathbf{h}}(\alpha)| \ll \frac{B^{2n}}{H^n} + \frac{B^n}{H^n} \sum_{\mathbf{h} \neq \mathbf{0}} |T_{\mathbf{h}}(\alpha)|,$$

where

$$T_{\mathbf{h}}(\alpha) := \sum_{\substack{\mathbf{x} \in \mathbb{Z}^n \\ |\mathbf{x}|, |\mathbf{x}+\mathbf{h}| \leqslant B}} e\big(\alpha(f(\mathbf{x}+\mathbf{h}) - f(\mathbf{x}))\big).$$

The reader should note that the special case $H = B$ of van der Corput's method reduces to the first step in Birch's approach.

For each non-zero $\mathbf{h} \in \mathbb{Z}^n$ the exponential sum $T_{\mathbf{h}}(\alpha)$ is a cubic exponential sum, involving the cubic polynomial $f_{\mathbf{h}}(\mathbf{x}) = f(\mathbf{x}+\mathbf{h}) - f(\mathbf{x})$. The cubic part of $f_{\mathbf{h}}(\mathbf{x})$ is equal to $\mathbf{h}.\nabla f(\mathbf{x})$. The idea in [34] is to estimate these exponential sums directly, using Poisson summation. This leads to the consideration of certain complete exponential sums modulo q, taking the shape

$$T_{\mathbf{h}}(a, q; \mathbf{v}) = \sum_{\mathbf{y} \pmod{q}} e_q(a f_{\mathbf{h}}(\mathbf{y}) + \mathbf{v}.\mathbf{y}),$$

for coprime a, q and $\mathbf{v} \in \mathbb{Z}^n$ restricted to some bounded region that expands with B. Each sum will satisfy a basic multiplicativity property that renders it sufficient to study the sums for prime power moduli p^ν, for each $p^\nu \| q$.

When $\nu = 1$, there are excellent bounds available for the sum coming from Deligne's estimates, provided that the singular locus of the projective hypersurface $\mathbf{h}.\nabla f(\mathbf{x}) = 0$ is not too large. Thus, in particular, one needs control over how frequently a given choice of $\mathbf{h} \in \mathbb{Z}^n$ produces a singular locus of large dimension, both as a variety over $\overline{\mathbb{Q}}$ and as a variety over $\overline{\mathbb{F}}_p$, for each $p \mid q$. When $\nu > 1$, one can no longer get satisfactory individual estimates for $T_{\mathbf{h}}(a, p^\nu; \mathbf{v})$. Instead one uses an elementary argument, which bounds an average of sums $T_{\mathbf{h}}(a, q; \mathbf{v})$, taken over a range of values for $\mathbf{v}$.

8.3 Diagonal cubic surfaces

We have seen in Chapter 2, and in particular Conjecture 2.3, that for any del Pezzo surface $S \subset \mathbb{P}^d$ one expects a growth rate like $c_S B (\log B)^A$ for the counting function $N_U(B)$. We have already given some motivation for the exponent of B in (1.3). The focus of the present section is to produce much more sophisticated heuristics than we have previously met. In particular we will gain an insight into the exponent of $\log B$ that appears in the Manin conjecture.

For ease of presentation we restrict attention to non-singular diagonal cubic surfaces $S \subset \mathbb{P}^3$. Thus

$$S = \{a_1 x_1^3 + a_2 x_2^3 + a_3 x_3^3 + a_4 x_4^3 = 0\}, \tag{8.26}$$

for $\mathbf{a} = (a_1, \dots, a_4) \in \mathbb{N}^4$ such that $\gcd(a_1, \dots, a_4) = 1$. Define

$$\mathscr{P} := \{3\} \cup \{p : p \mid a_1a_2a_3a_4\}. \tag{8.27}$$

This is the set of primes p for which the reduction of S modulo p is singular. Before passing to the arithmetic of diagonal cubic surfaces, and our application of the Hardy–Littlewood circle method, we will need to discuss some of their geometry.

8.3.1 The lines on a diagonal cubic surface

In this section we expand upon the calculations made in Section 2.1.4, concerning the 27 lines on a cubic surface. We focus our attention upon surfaces S of the shape (8.26). In this setting it is not hard to write down the 27 lines explicitly.

The calculations that we present below are based on those carried out by Peyre and Tschinkel [99, Section 2]. Fix a cubic root α (resp. α', α'') of $\frac{a_2}{a_1}$ (resp. $\frac{a_3}{a_1}, \frac{a_4}{a_1}$). We will assume that $\alpha \in \mathbb{Q}$ (resp. $\alpha', \alpha'' \in \mathbb{Q}$) if $\frac{a_2}{a_1}$ (resp. $\frac{a_3}{a_1}, \frac{a_4}{a_1}$) is a cube in $\mathbb{Q}$. Put

$$\beta = \frac{\alpha''}{\alpha'}, \quad \beta' = \frac{\alpha}{\alpha''}, \quad \beta'' = \frac{\alpha'}{\alpha}.$$

We denote by θ a primitive cube root of 1. Let i run over elements of $\mathbb{Z}/3\mathbb{Z}$. Then the 27 lines on the cubic surface (8.26) are given by the equations

$$L_i{:}\begin{cases} x_1+\theta^i\alpha x_2 = 0,\\ x_3+\theta^i\beta x_4 = 0,\end{cases} \quad L_i'{:}\begin{cases} x_1+\theta^i\alpha x_2 = 0,\\ x_3+\theta^{i+1}\beta x_4 = 0,\end{cases} \quad L_i''{:}\begin{cases} x_1+\theta^i\alpha x_2 = 0,\\ x_3+\theta^{i+2}\beta x_4 = 0,\end{cases}$$

$$M_i{:}\begin{cases} x_1+\theta^i\alpha' x_3 = 0,\\ x_4+\theta^i\beta' x_2 = 0,\end{cases} \quad M_i'{:}\begin{cases} x_1+\theta^i\alpha' x_3 = 0,\\ x_4+\theta^{i+1}\beta' x_2 = 0,\end{cases} \quad M_i''{:}\begin{cases} x_1+\theta^i\alpha' x_3 = 0,\\ x_4+\theta^{i+2}\beta' x_2 = 0,\end{cases}$$

$$N_i{:}\begin{cases} x_1+\theta^i\alpha'' x_4 = 0,\\ x_2+\theta^i\beta'' x_3 = 0,\end{cases} \quad N_i'{:}\begin{cases} x_1+\theta^i\alpha'' x_4 = 0,\\ x_2+\theta^{i+1}\beta'' x_3 = 0,\end{cases} \quad N_i''{:}\begin{cases} x_1+\theta^i\alpha'' x_4 = 0,\\ x_2+\theta^{i+2}\beta'' x_3 = 0.\end{cases}$$

Let $K = \mathbb{Q}(\theta, \alpha, \alpha', \alpha'')$. It is a Galois extension of $\mathbb{Q}$, and in the generic case has degree 54 with Galois group $\mathrm{Gal}(K/\mathbb{Q}) \cong (\mathbb{Z}/3\mathbb{Z})^3 \rtimes \mathbb{Z}/2\mathbb{Z}$.

We need to equate these lines to the divisors $E_i, L_{i,j}, Q_i$ that we met in Section 2.1.4. There is a certain degree of freedom in doing this, as discussed in

[66, Section V.4], but it turns out that the choice

$$\begin{array}{lll} E_1 = L_0, & E_2 = L_1, & E_3 = L_2, \\ E_4 = M_1, & E_5 = M_2', & E_6 = M_0'', \\ Q_1 = L_1', & Q_2 = L_2', & Q_3 = L_0', \\ Q_4 = M_0, & Q_5 = M_1', & Q_6 = M_2'', \\ L_{1,2} = L_1'', & L_{2,3} = L_2'', & L_{1,3} = L_0'', \\ L_{4,5} = M_1'', & L_{5,6} = M_2, & L_{4,6} = M_0', \\ L_{1,4} = N_0, & L_{1,5} = N_1, & L_{1,6} = N_2, \\ L_{2,4} = N_1', & L_{2,5} = N_2', & L_{2,6} = N_0', \\ L_{3,4} = N_2'', & L_{3,5} = N_0'', & L_{3,6} = N_1'', \end{array}$$

is satisfactory. In assigning lines to $E_1, \dots, E_6$, all that is required is that they should be mutually skew and globally invariant under the action of the Galois group. Given *any* cubic surface of the shape (8.26), we now have the tools with which to compute the Picard group (2.7). In fact, from this point forwards the process requires little more than linear algebra.

Let us illustrate the procedure by calculating the Picard group for a special case. Consider the Fermat surface

$$S_1 = \{x_1^3 + x_2^3 + x_3^3 + x_4^3 = 0\}. \tag{8.28}$$

In this case $\alpha = \alpha' = \alpha'' = \beta = \beta' = \beta'' = 1$, in the notation above, and $K = \mathbb{Q}(\theta)$ is a quadratic field extension. We wish to find elements of the geometric Picard group $\mathrm{Pic}_{\overline{\mathbb{Q}}}(S_1)$ that are fixed by the action of $\mathrm{Gal}(K/\mathbb{Q})$. Recall that $\mathrm{Pic}_{\overline{\mathbb{Q}}}(S_1)$ is generated by $[\Lambda], [E_1], \dots, [E_6]$. Thus we want vectors $\mathbf{c} = (c_0, \dots, c_6) \in \mathbb{Z}^7$ such that

$$(c_0[\Lambda] + c_1[E_1] + \dots + c_6[E_6])^\sigma = c_0[\Lambda] + c_1[E_1] + \dots + c_6[E_6], \tag{8.29}$$

for every $\sigma \in \mathrm{Gal}(K/\mathbb{Q})$. Under the action of $\mathrm{Gal}(K/\mathbb{Q})$ it is not hard to check that Λ and E_1 are fixed, that E_2 and E_3 are swapped, and that E_4 (resp. E_5, E_6) is taken to $L_{5,6}$ (resp. $L_{4,5}$, $L_{4,6}$). Using (2.3) one sees that the left-hand side of (8.29) is equal to

$$\begin{aligned}(c_0 + c_4 + c_5 + c_6)[\Lambda] + & c_1[E_1] + c_2[E_3] + c_3[E_2] \\ & - (c_5 + c_6)[E_4] - (c_4 + c_5)[E_5] - (c_4 + c_6)[E_6],\end{aligned}$$

in $\mathrm{Pic}_{\overline{\mathbb{Q}}}(S_1)$. Thus we are interested in the space of $\mathbf{c} \in \mathbb{Z}^7$ for which

$$c_4 + c_5 + c_6 = 0, \quad c_2 - c_3 = 0, \quad c_4 + 2c_5 = 0, \quad c_4 + 2c_6 = 0.$$

This system of homogeneous linear equations in 7 variables has an underlying matrix of rank 3. Thus the space of solutions has rank $7 - 3 = 4$, and so we may conclude that $\mathrm{Pic}(S_1) \cong \mathbb{Z}^4$. In fact a little thought reveals that the 4 elements

$$[\Lambda], [E_1], [E_2] + [E_3], -2[E_4] + [E_5] + [E_6]$$

provide a basis for $\mathrm{Pic}(S_1)$.

Next, consider the surface

$$S_2^{(p)} = \{x_1^3 + x_2^3 + x_3^3 + px_4^3 = 0\} \tag{8.30}$$

for a prime p. When $p = 2$ or 3, the arithmetic of this surface has been considered in some detail by Heath-Brown [69], who provides some numerical evidence to the effect that the corresponding counting function $N_U(B)$ should grow like c_pB for a certain constant $c_p > 0$. This is extended considerably by Peyre and Tschinkel [98]. By Exercise 8.2 we have $\mathrm{Pic}(S_2^{(p)}) \cong \mathbb{Z}$. More generally, it is known that the Picard group of the surface (8.26) has rank 1 if and only if the ratio

$$\frac{a_{\sigma(1)}a_{\sigma(2)}}{a_{\sigma(3)}a_{\sigma(4)}}$$

is not a cube in $\mathbb{Q}$, for each permutation σ of $(1, 2, 3, 4)$. This is due to Segre [109].

8.3.2 Cubic characters and Jacobi sums

Throughout this section let p be a prime. A (multiplicative) *character* on $\mathbb{F}_p$ is a map $\chi : \mathbb{F}_p^* \to \mathbb{C}^*$ such that

$$\chi(ab) = \chi(a)\chi(b)$$

for all $a, b \in \mathbb{F}_p^*$. The *trivial character* ε is defined by the relation $\varepsilon(a) = 1$ for all $a \in \mathbb{F}_p^*$. It is convenient to extend the domain of definition to all of $\mathbb{F}_p$ by assigning $\chi(0) = 0$ if $\chi \neq \varepsilon$ and $\varepsilon(0) = 1$.

We begin by collecting together a few basic facts, all of which are established in the book by Ireland and Rosen [84, Section 8].

Lemma 8.9. *Let p be a prime. Then the following hold:*

(i) *for any character χ on $\mathbb{F}_p$ and any $a \in \mathbb{F}_p^*$, we have $\chi(1) = 1$ and $\chi(a)$ is a $(p-1)$th root of unity, and furthermore, $\chi(a^{-1}) = \overline{\chi}(a) = \chi(a)^{-1}$;*

(ii) *for any character χ on $\mathbb{F}_p$ we have*

$$\sum_{a\in\mathbb{F}_p} \chi(a) = \begin{cases} 0, & \text{if } \chi \neq \varepsilon, \\ p, & \text{if } \chi = \varepsilon; \end{cases}$$

(iii) *the set of characters on $\mathbb{F}_p$ forms a cyclic group of order $p - 1$.*

It follows from part (iii) of Lemma 8.9 that $\chi^{p-1} = \varepsilon$ for any character on $\mathbb{F}_p$. We define the *order* of a character to be the least positive integer n such that $\chi^n = \varepsilon$. In our work we will mainly be concerned with the characters of order 3.

Let us turn briefly to the topic of generalised Jacobi sums. Given any characters $\chi_1, \ldots, \chi_r$ on $\mathbb{F}_p$, a *Jacobi sum* is a sum of the shape

$$J_0(\chi_1, \ldots, \chi_r) := \sum_{\substack{\mathbf{t}=(t_1,\ldots,t_r)\in\mathbb{F}_p^r \\ t_1+\cdots+t_r\equiv 0 \pmod p}} \chi_1(t_1)\cdots\chi_r(t_r).$$

The key fact that we will need concerning these sums is that

$$|J_0(\chi_1, \ldots, \chi_r)| = \begin{cases} 0, & \text{if } \chi_1\cdots\chi_r \neq \varepsilon, \\ (p-1)p^{\frac{r}{2}-1}, & \text{if } \chi_1\cdots\chi_r = \varepsilon. \end{cases} \tag{8.31}$$

This is established in [84, Section 8.5].

Let p be a rational prime. We proceed to consider p as an element of the ring of integers $\mathbb{Z}[\theta]$ associated to the quadratic field $\mathbb{Q}(\theta)$ obtained by adjoining a primitive cube root of unity θ. It follows that p is a prime in $\mathbb{Z}[\theta]$ if $p \equiv 2 \pmod 3$, whereas it splits as $p = \pi\overline{\pi}$ if $p \equiv 1 \pmod 3$, where π is a prime in $\mathbb{Z}[\theta]$. When $p \equiv 2 \pmod 3$ the only cubic character on $\mathbb{F}_p$ is the trivial character ε. On the other hand, when $p = \pi\overline{\pi} \equiv 1 \pmod 3$, then there are precisely two non-trivial cubic characters $\chi_\pi, \chi_{\overline{\pi}}$ on $\mathbb{F}_p$, where

$$\chi_\omega(\cdot) = \left(\frac{\cdot}{\omega}\right)_3$$

is the cubic residue symbol for any prime ω in $\mathbb{Z}[\theta]$. All of these facts are established in [84, Section 9].

It turns out that Jacobi sums can be used to give formulae for the number of solutions to diagonal equations over finite fields. This point of view was crucial in Weil's [121] resolution of the Weil conjectures for diagonal hypersurfaces. Given any $q \in \mathbb{N}$, let

$$N(q) := \#\{\mathbf{x} \pmod q : a_1x_1^3 + \cdots + a_4x_4^3 \equiv 0 \pmod q\}, \tag{8.32}$$

and

$$N^*(q) := \#\left\{\mathbf{x} \pmod q : \begin{array}{l} a_1x_1^3 + \cdots + a_4x_4^3 \equiv 0 \pmod q, \\ \gcd(q, x_1, \ldots, x_4) = 1 \end{array}\right\}. \tag{8.33}$$

When p is a prime not belonging to the finite set of primes $\mathscr{P}$ defined in (8.27), we can write down a very precise expression for $N(p)$. Thus it follows from [84, Section 8.7] that

$$N(p) = p^3 + \sum_{\chi_1,\chi_2,\chi_3,\chi_4} \chi_1(a_1^{-1})\chi_2(a_2^{-1})\chi_3(a_3^{-1})\chi_4(a_4^{-1})J_0(\chi_1,\chi_2,\chi_3,\chi_4)$$

where the summation is over all non-trivial cubic characters $\chi_i : \mathbb{F}_p^* \to \mathbb{C}$ such that $\chi_1\chi_2\chi_3\chi_4 = \varepsilon$.

It follows from Exercise 8.3 that

$$N^*(p) = N(p) - 1 = p^3 + p(p-1)\delta_p(\mathbf{a}) - 1, \tag{8.34}$$

for any prime $p \notin \mathscr{P}$, where

$$\delta_p(\mathbf{a}) := \sum_{\chi_1,\chi_2,\chi_3,\chi_4} \chi_1(a_1^{-1})\chi_2(a_2^{-1})\chi_3(a_3^{-1})\chi_4(a_4^{-1}),$$

and the summation is over all non-trivial cubic characters such that $\chi_1 \cdots \chi_4 = \varepsilon$. Let $a \in \mathbb{F}_p^*$ and suppose that p splits as $\pi\overline{\pi}$. Then it will be useful to observe that

$$\chi_\pi(a) + \chi_{\overline{\pi}}(a) = \begin{cases} 2, & \text{if } a \text{ is a cubic residue modulo } \pi, \\ -1, & \text{otherwise.} \end{cases} \tag{8.35}$$

We have $\delta_p(\mathbf{a}) = 0$ when $p \equiv 2 \pmod 3$, since there are then no non-trivial cubic characters modulo p. When $p \equiv 1 \pmod 3$, with $p = \pi\overline{\pi} \notin \mathscr{P}$, we have

$$\begin{aligned}\delta_p(\mathbf{a}) = &\chi_\pi\Big(\frac{a_1a_2}{a_3a_4}\Big) + \chi_{\overline{\pi}}\Big(\frac{a_1a_2}{a_3a_4}\Big) + \chi_\pi\Big(\frac{a_1a_3}{a_2a_4}\Big) + \chi_{\overline{\pi}}\Big(\frac{a_1a_3}{a_2a_4}\Big) \\ &+ \chi_\pi\Big(\frac{a_1a_4}{a_2a_3}\Big) + \chi_{\overline{\pi}}\Big(\frac{a_1a_4}{a_2a_3}\Big).\end{aligned}$$

Still with this choice of prime p, let us write $\nu_p(\mathbf{a})$ for the number of indices $i \in \{2,3,4\}$ such that the cubic character $\chi_\pi(\frac{a_1a_i}{a_ja_k})$ is equal to 1, with $\{i,j,k\}$ a permutation of $\{2,3,4\}$. Then (8.35) implies that

$$\delta_p(\mathbf{a}) = \begin{cases} 0, & \text{if } p \equiv 2 \pmod 3, \\ 3\nu_p(\mathbf{a}) - 3, & \text{if } p \equiv 1 \pmod 3, \end{cases} \tag{8.36}$$

when $p \notin \mathscr{P}$.

8.3.3 The heuristic

We are now ready to consider the counting function $N_U(B)$ that is associated to the diagonal cubic surface $S \subset \mathbb{P}^3$ given in (8.26). Our aim is to provide heuristic evidence in support of Manin's original conjecture, and we will say rather little about the predicted value of the constant. The key will be to apply the Hardy–Littlewood method, but only to consider the contribution from the major arcs.

We will simplify matters by applying the heuristic to count all of the rational points on S, rather than restricting attention to the open subset U. Although the details are formidable, it is in fact possible to obtain upper bounds for $N_U(B)$ using the circle method. Thus Heath-Brown [72] has shown that $N_U(B) = O_{\varepsilon,S}(B^{\frac{3}{2}+\varepsilon})$ under a certain hypothesis concerning the Hasse–Weil L-function associated to the surface. An interesting feature of this work is that the contribution from the rational points lying on rational lines in the surface is successfully separated out. When the surface contains no lines defined over $\mathbb{Q}$, such as the surface given by (8.30) for example, one obviously has

$$N_U(B) = N_S(B) + O(1).$$

When S contains lines defined over $\mathbb{Q}$, there is a general consensus among people working on the circle method that the dominant contribution (i.e., the contribution from the points on rational lines) should come from the minor arc integral.

Applying Möbius inversion, we deduce from (8.1) that

$$\begin{aligned} N_S(B) &= \frac{1}{2}\int_0^1 \sum_{\substack{\mathbf{x}\in\mathbb{Z}^4_{\text{prim}}\\ |\mathbf{x}|\leqslant B}} e(\alpha(a_1x_1^3+\cdots+a_4x_4^3))\mathrm{d}\alpha \\ &= \frac{1}{2}\sum_{k=1}^{\infty}\mu(k)\int_0^1 \sum_{\substack{\mathbf{x}\in\mathbb{Z}^4\\ |\mathbf{x}|\leqslant k^{-1}B}} e(\alpha(a_1x_1^3+\cdots+a_4x_4^3))\mathrm{d}\alpha \\ &= \frac{1}{2}\sum_{k=1}^{\infty}\mu(k)\int_0^1 S(\alpha)\mathrm{d}\alpha, \end{aligned}$$

say. Let us write $P = k^{-1}B$ and $I_{\mathscr{A}}(P) := \int_{\mathscr{A}} S(\alpha)\mathrm{d}\alpha$, for any $\mathscr{A}\subset\mathbb{R}$.

For the purposes of our heuristic we will analyse $I_{\mathfrak{M}(\varepsilon)}(P)$ for the set of major arcs defined in (8.7) with $d=3$. Here $\varepsilon>0$ is a small positive constant. In particular Lemma 8.3 ensures that the arcs are all disjoint. We will ignore the contribution from the minor arc integral $I_{\mathfrak{m}(\varepsilon)}(P)$. In truth there will be several points in the argument where we will simply ignore subsidiary contributions. We will indicate all of these by an appearance of the word "error". Thus, to begin with, we have

$$N_U(B) = \frac{1}{2}\sum_{k=1}^{\infty}\mu(k)I_{\mathfrak{M}(\varepsilon)}(P) + \text{error}.$$

Here we have made the further assumption that the contribution $N_{S\setminus U}(B)$ from the points lying on lines in S arises in the minor arc integral.

Let $\alpha = \frac{a}{q}+z \in \mathfrak{M}_{q,a}(\varepsilon)$ for given coprimes $a,q\in\mathbb{N}$ such that $a\leqslant q\leqslant P^{\varepsilon}$. Let $f(\mathbf{x})$ denote the diagonal cubic form in (8.26). It follows from Lemma 8.2 that

$$S\Big(\frac{a}{q}+z\Big) = q^{-4}B^4 S_{q,a} I(zP^3) + O(P^{3+2\varepsilon}),$$

where the $S_{q,a}, I(\gamma)$ are given by (8.2) and (8.6), respectively.

The set of major arcs has measure $O(P^{-3+3\varepsilon})$. On carrying out the integration over z and the summation over a and q one is therefore led to the conclusion that

$$\begin{aligned}
I_{\mathfrak{M}(\varepsilon)}(P) &= P^4 \sum_{q\leqslant P^\varepsilon} \sum_{\substack{1\leqslant a\leqslant q\\ \gcd(a,q)=1}} q^{-4} S_{q,a} \int_{|z|\leqslant P^{-3+\varepsilon}} I(zP^3)\mathrm{d}z + O(P^{5\varepsilon})\\
&= \mathfrak{S}(P^\varepsilon)\mathfrak{I}(P^\varepsilon)P \sum_{q\leqslant P^\varepsilon} \sum_{\substack{1\leqslant a\leqslant q\\ \gcd(a,q)=1}} q^{-4} S_{q,a} + O(P^{5\varepsilon}),
\end{aligned}$$

in the notation of (8.21) and (8.22), with $n = 4$. It can be shown that $\mathfrak{I}(R)$ is a bounded function of R, and furthermore, $\mathfrak{I}(R) \to \mathfrak{I} > 0$ as $R \to \infty$. A standard calculation reveals that $\mathfrak{I} = 2\sigma_\infty$, where

$$\sigma_\infty := \frac{1}{6a_4^{\frac{1}{3}}} \int \frac{\mathrm{d}x_1\mathrm{d}x_2\mathrm{d}x_3}{(a_1x_1^3 + a_2x_2^3 + a_3x_3^3)^{\frac{2}{3}}}$$

is the real density of solutions. Here the integral is over $x_1, x_2, x_3 \in [-1, 1]$ such that $|a_4^{-1}(a_1x_1^3 + a_2x_2^3 + a_3x_3^3)| \leqslant 1$. On bringing everything together, our investigation has so far succeeded in showing that

$$N_U(B) = \sigma_\infty \sum_{k=1}^{\infty} \mu(k)P\mathfrak{S}(P^\varepsilon) + \text{error}, \tag{8.37}$$

for a suitably small value of $\varepsilon > 0$, where $P = k^{-1}B$.

Our task is now to examine the sum $\mathfrak{S}(R)$, as $R \to \infty$. Let us write

$$S_q := \sum_{\substack{1\leqslant a\leqslant q\\ \gcd(a,q)=1}} S_{q,a}. \tag{8.38}$$

Then $\mathfrak{S}(R) = \sum_{q\leqslant R} q^{-4}S_q$. It turns out that S_q is a multiplicative function of q. This can be established along the lines of [50, Lemma 5.1]. Recall the definition (8.32) of $N(q)$. We now come to the key relation between S_q and $N(q)$ at prime power values of q. Indeed, it follows from Exercise 8.4 that

$$S_{p^e} = p^e N(p^e) - p^{3+e}N(p^{e-1}),$$

for any prime p and any $e \in \mathbb{N}$.

Let us for the moment ignore issues of convergence, and consider the local factors $\sum_{e=0}^{\infty} p^{-4e}S_{p^e}$ in the infinite product formula for $\mathfrak{S}$. Now it follows from the above that

$$\sum_{e=0}^{E} p^{-4e}S_{p^e} = 1 + \sum_{e=1}^{E} (p^{-3e}N(p^e) - p^{3-3e}N(p^{e-1})) = p^{-3E}N(p^E),$$

for any $E \geqslant 1$. Hence, formally speaking, we have $\mathfrak{S} = \prod_p \tau_p$, where

$$\tau_p := \lim_{e\to\infty} p^{-3e} N(p^e).$$

If $\mathfrak{S}(R)$ was absolutely convergent, which it certainly is not in general, we could then conclude from (8.37) that

$$N_U(B) = B\sigma_\infty \sum_{k=1}^{\infty} \frac{\mu(k)}{k} \prod_p \tau_p + \text{error}. \tag{8.39}$$

Arguing formally, we now replace the summation over k by its Euler product, concluding that

$$N_U(B) = B\sigma_\infty \prod_p \sigma_p + \text{error}, \tag{8.40}$$

with $\sigma_p := (1 - \frac{1}{p})\tau_p$. Note that

$$\sigma_p = \lim_{e\to\infty} p^{-3e} N^*(p^e), \tag{8.41}$$

in the notation of (8.33), which follows from the observation that

$$N^*(p^e) = N(p^e) - p^8 N(p^{e-3}),$$

for any $e > 3$. The estimate in (8.40) therefore gives a heuristic asymptotic formula for $N_U(B)$ which is visibly a product of local densities. Among other things, we have assumed that $\mathfrak{S}(R)$ is absolutely convergent in formulating this heuristic.

One can make the transition from (8.39) to (8.40) completely rigorous if the Hardy–Littlewood heuristic produces a leading term involving P^α with exponent $\alpha > 1$. The outcome is that to go from counting points on the affine cone to counting projective points, one merely replaces $N(p^e)$ by $N^*(p^e)$. For $\alpha = 1$, however, the *renormalisation* procedure remains heuristic. Let

$$\mathfrak{S}^*(R) := \sum_{q\leqslant R} q^{-4} S_q^*,$$

where

$$S_q^* := \sum_{\substack{1\leqslant a\leqslant q\\ \gcd(a,q)=1}} \sum_{\substack{\mathbf{r} \pmod q\\ \gcd(\mathbf{r},q)=1}} e_q(af(\mathbf{r})).$$

It is easily checked that S_q^* is a multiplicative function of q, and furthermore, that the corresponding version of Exercise 8.4 holds, relating $S_{p^e}^*$ to $N^*(p^e)$. In fact, formally speaking, one has

$$\prod_p \sum_{e=0}^{\infty} p^{-4e} S_{p^e}^* = \prod_p \sigma_p,$$

with σ_p given by (8.41). Bearing all of this in mind we will proceed under the bold assumption that (8.37) can be replaced by

$$N_U(B) = B\sigma_\infty \mathfrak{S}^*(B) + \text{error}, \tag{8.42}$$

where we have taken $\varepsilon = 1$ in the expressions for $\mathfrak{S}(B^\varepsilon)$ and $\mathfrak{S}^*(B^\varepsilon)$.

We now turn to a finer analysis of $\mathfrak{S}^*(B)$, as $B \to \infty$. Our task is to determine the analytic properties of the corresponding Dirichlet series

$$F(s) := \sum_{q=1}^{\infty} \frac{S_q^*}{q^s}$$

for $s = \sigma + it \in \mathbb{C}$. Armed with this analysis we will ultimately apply Perron's formula to obtain an estimate for $\mathfrak{S}^*(B)$. Using the multiplicativity of $q^{-s}S_q^*$ we deduce that

$$F(s) = \prod_p \sigma_p(s), \quad \sigma_p(s) := \sum_{e=0}^{\infty} p^{-es} S_{p^e}^*.$$

In examining $F(s)$ it clearly suffices to ignore the value of the factors $\sigma_p(s)$ at any finite collection of primes p. With this in mind we will try and determine $\sigma_p(s)$ for $p \notin \mathscr{P}$, where $\mathscr{P}$ is given by (8.27). Recall the definition (8.33) of $N^*(q)$. It therefore follows from Exercise 8.5 that

$$\sigma_p(s) = 1 + \sum_{e=1}^{\infty} p^{-es}(p^e N^*(p^e) - p^{3+e}N^*(p^{e-1})) = 1 - \frac{1}{p^{s-4}} + \frac{N^*(p)}{p^{s-1}},$$

for any $p \notin \mathscr{P}$. Hence (8.34) yields

$$\sigma_p(s) = 1 + \frac{\delta_p(\mathbf{a})}{p^{s-3}} - \frac{\delta_p(\mathbf{a})}{p^{s-2}} - \frac{1}{p^{s-1}},$$

where $\delta_p(\mathbf{a})$ is given by (8.36).

We now pursue our analysis in the special case $\mathbf{a} = (1,1,1,1)$ of the Fermat cubic surface (8.28). It is clear from (8.36) that $\delta_p(1,1,1,1) = 0$ if $p \equiv 2 \pmod 3$ and

$$\delta_p(1,1,1,1) = 3\nu_p(1,1,1,1) - 3 = 6$$

if $p \equiv 1 \pmod 3$. Let $\lambda : \mathbb{Z} \to \mathbb{C}$ be the real Dirichlet character of order 2 defined by

$$\lambda(n) := \begin{cases} (\frac{n}{3}), & \text{if } 3 \nmid n, \\ 0, & \text{otherwise,} \end{cases}$$

where $(\frac{n}{3})$ is the Legendre symbol. Then we may write

$$\begin{aligned} \sigma_p(s) &= 1 + \frac{3(1+\lambda(p))}{p^{s-3}} - \frac{3(1+\lambda(p))}{p^{s-2}} - \frac{1}{p^{s-1}} \\ &= \Big(1 - \frac{1}{p^{s-3}}\Big)^{-3}\Big(1 - \frac{\lambda(p)}{p^{s-3}}\Big)^{-3}\Big(1 + O\Big(\frac{1}{p^{\min\{\sigma-2, 2\sigma-6\}}}\Big)\Big), \end{aligned}$$

for any $p \notin \mathscr{P}$. Let $L(s,\lambda)$ denote the usual Dirichlet L-function associated to λ. When $\mathbf{a} = (1,1,1,1)$, we have therefore succeeded in showing that

$$F(s) = \zeta(s-3)^3 L(s-3,\lambda)^3 G(s), \tag{8.43}$$

where $G(s)$ is holomorphic and bounded on the half-plane $\sigma \geqslant \frac{7}{2}+\delta$, for any $\delta > 0$. For future reference we note that $G(4)$ has local factors

$$G_p(4) = \begin{cases} (1-\frac{1}{p})^7(1+\frac{7}{p}+\frac{1}{p^2}), & \text{if } p \equiv 1 \pmod 3, \\ (1-\frac{1}{p})^4(1+\frac{1}{p})^3(1+\frac{1}{p}+\frac{1}{p^2}), & \text{if } p \equiv 2 \pmod 3. \end{cases} \tag{8.44}$$

Although we will not prove it here, it can be deduced from (8.41) and Hensel's lemma that

$$G_3(4) = \frac{8}{27}\lim_{e\to\infty} 3^{-3e} N^*(3^e) = \frac{16}{27}. \tag{8.45}$$

We are now ready for our application of Perron's formula, in the form of Lemma 5.4. Clearly $F(s+4)$ has abscissa of absolute convergence 0. Taking $c=2$ and $B \notin \mathbb{Z}$, we easily deduce that

$$\mathfrak{S}^*(B) = \frac{1}{2\pi i}\int_{2-iT}^{2+iT} F(s+4)\frac{B^s}{s}\mathrm{d}s + O\Big(\frac{B^2\log B}{T}\Big).$$

Here the contribution from the error terms in Lemma 5.4 is handled in a similar fashion to our application of Perron's formula in Section 5.3.

We now apply Cauchy's residue theorem to the rectangular contour $\mathscr{C}$ joining $c'-iT$, $c'+iT$, $2+iT$ and $2-iT$, where $c' = -\frac{1}{2}+\varepsilon$. The relation (8.43) implies that in this region $F(s+4)\frac{B^s}{s}$ has a unique pole at $s=0$, and it is a pole of order 4. It has residue

$$\mathrm{Res}_{s=0}\frac{F(s+4)B^s}{s} = \frac{L(1,\lambda)^3 G(4) P(\log B)}{3!},$$

where $P \in \mathbb{R}[x]$ is a monic polynomial of degree 3. Putting all of this together we have therefore shown that

$$\mathfrak{S}^*(B) = \frac{L(1,\lambda)^3 G(4) P(\log B)}{3!} + O(E(B)), \tag{8.46}$$

where

$$E(B) = \frac{B^2\log B}{T} + \Big(\int_{c'-iT}^{c'+iT} + \int_{c'-iT}^{2-iT} + \int_{2+iT}^{c'+iT}\Big)\Big|H(s)^3\frac{B^s}{s}\Big|\mathrm{d}s,$$

for any $T \geqslant 1$, with $H(s) = \zeta(s+1)L(s+1,\lambda)$. Here we have used the fact that $G(s+4)$ is bounded on the half-plane $\Re e(s) \geqslant c'$.

To make the analysis simpler, it will be convenient to proceed under the assumption that the *Lindelöf hypothesis* holds for $\zeta(s)$, and also for the Dirichlet

L-function $L(s,\lambda)$. This could be avoided at the cost of extra effort, but there is no harm in supposing it here. Thus we may assume the bounds

$$\zeta(\sigma+it) \ll_\varepsilon |t|^\varepsilon, \quad L(\sigma+it,\lambda) \ll_\varepsilon |t|^\varepsilon,$$

for any $\sigma \in [\frac{1}{2},1]$ and any $|t| \geqslant 1$. It therefore follows that

$$H(\sigma+it) \ll_\varepsilon \begin{cases} |t|^\varepsilon, & \text{if } -\frac{1}{2} \leqslant \sigma \leqslant 0, \\ 1, & \text{if } \sigma > 0, \end{cases}$$

for any $\varepsilon > 0$ and $|t| \geqslant 1$. This gives

$$\int_{c'-iT}^{2-iT} \left| H(s)^3 \frac{B^s}{s} \right| \mathrm{d}s \ll_\varepsilon \int_{c'}^{2} B^\sigma T^{-1+3\varepsilon} \mathrm{d}\sigma \ll_\varepsilon \frac{B^2 T^{3\varepsilon}}{T}.$$

One obtains the same estimate for the contribution from the remaining horizontal contour. Turning to vertical integral, we find that

$$\begin{aligned} \int_{c'-iT}^{c'+iT} \left| H(s)^3 \frac{B^s}{s} \right| \mathrm{d}s &\ll B^{c'} \int_{-T}^{T} \frac{|H(-\frac{1}{2}+\varepsilon+it)|^3}{1+|t|} \mathrm{d}t \\ &\ll_\varepsilon B^{c'} \int_{-T}^{T} (1+|t|)^{3\varepsilon-1} \mathrm{d}t \\ &\ll_\varepsilon B^{c'} T^{3\varepsilon}. \end{aligned}$$

This shows that $E(B) \ll_\varepsilon B^\varepsilon T^{3\varepsilon}(B^2 T^{-1} + B^{-\frac{1}{2}})$, for any $T \geqslant 1$. Taking T sufficiently large, we therefore conclude from (8.46) that

$$\mathfrak{S}^*(B) = \frac{L(1,\lambda)^3 G(4) P(\log B)}{3!} + O(B^{-\Delta}),$$

for some $\Delta > 0$.

We are now ready to return to the Hardy–Littlewood major arc analysis which led us to (8.42). Substituting in our estimate for $\mathfrak{S}^*(B)$, we conclude that

$$N_{U_1}(B) = c_1 B (\log B)^3 + \text{error}, \tag{8.47}$$

where $U_1 \subset S_1$ is the usual open subset of the Fermat surface (8.28) and

$$c_1 = \frac{\sigma_\infty L(1,\lambda)^3 G(4)}{3!}.$$

Now it follows from the class number formula that $L(1,\lambda) = \frac{\pi\sqrt{3}}{9}$. Hence, on combining this with (8.44) and (8.45), our heuristic argument has led us to the expectation that (8.47) holds, with

$$c_1 = \frac{\sigma_\infty 2^4 \pi^3 \sqrt{3}}{3! 3^8} \prod_{p\equiv 1 \ (\mathrm{mod}\ 3)} \left(1-\frac{1}{p}\right)^7 \left(1+\frac{7}{p}+\frac{1}{p^2}\right) \prod_{p\equiv 2 \ (\mathrm{mod}\ 3)} \left(1-\frac{1}{p^3}\right)\left(1-\frac{1}{p^2}\right)^3.$$

The exponents of B and $\log B$ in (8.47) agree with the Manin conjecture, since we have already seen that the Picard group of S_1 has rank 4.

It is interesting to compare our analysis with that of Peyre and Tschinkel [99], who calculate the leading constant c_{Peyre} in Peyre's refinement [96] of the conjectured asymptotic formula for $N_{U_1}(B)$. It turns out that

$$c_{\mathrm{Peyre}} = \gamma(S_1)c_1,$$

with $\gamma(S_1) = \frac{7}{3}$. For a general non-singular cubic surface $S \subset \mathbb{P}^3$, the constant $\gamma(S)$ is defined to be the volume

$$\gamma(S) := \int_{\Lambda_{\mathrm{eff}}^\vee(S)} e^{-\langle -K_S, \mathbf{t}\rangle} \mathrm{d}\mathbf{t}.$$

Thus, in general terms, $\gamma(S)$ measures the volume of the polytope obtained by intersecting the dual of $\Lambda_{\mathrm{eff}}(S)$ with a certain affine hyperplane. In particular $\gamma(S) \in \mathbb{Q}$ for any non-singular cubic surface S, and $\gamma(S) = 1$ if and only if the corresponding Picard group has rank 1.

Exercises for Chapter 8

Exercise 8.1. Let $f \in \mathbb{Z}[x_1, \ldots, x_n]$ be a non-singular quartic form. Assuming the estimate $\mathfrak{S}(R) = \mathfrak{S} + O_\delta(R^{-\delta})$, for some $\delta > 0$, show that the treatment of the major arcs in Section 8.2.2 can be handled for $n > 24$.

Exercise 8.2. Show that $\mathrm{Pic}(S_2^{(p)}) \cong \mathbb{Z}$, where $S_2^{(p)}$ is given by (8.30) for a prime number p.

Exercise 8.3. Let $p \neq 3$ be a prime, and let $\chi_1, \chi_2, \chi_3, \chi_4$ be non-trivial cubic characters on $\mathbb{F}_p$ such that $\chi_1\chi_2\chi_3\chi_4 = \varepsilon$. Deduce from (8.31) that

$$J_0(\chi_1, \chi_2, \chi_3, \chi_4) = p(p-1).$$

Exercise 8.4. Let p be a prime and let $e \geqslant 1$. Use Lemma 8.9 to show that

$$S_{p^e} = p^e N(p^e) - p^{3+e} N(p^{e-1}),$$

where $S_q, N(q)$ are given by (8.38) and (8.32), respectively.

Exercise 8.5. Recall the definition (8.33) of $N^*(q)$. Let $e \geqslant 1$ and let $p \notin \mathscr{P}$ be a prime, with $\mathscr{P}$ given by (8.27). Use Hensel's lemma to show that $N^*(p^e) = p^{3e-3}N^*(p)$.

Exercise 8.6. Let S_2 denote the surface (8.30) with $p = 2$. Using a similar argument, show that one expects an asymptotic formula of the shape

$$N_{S_2}(B) \sim c_2 B$$

for some constant $c_2 \geqslant 0$. Check your answer with the heuristic formula obtained by Heath-Brown [69].

Bibliography

[1] J. Ax and S. Kochen, Diophantine problems over local fields. I. *Amer. J. Math.* **87** (1965), 605–630.

[2] V.V. Batyrev and Yu.I. Manin, Sur le nombre des points rationnels de hauteur borné des variétés algébriques. *Math. Annalen* **286** (1990), 27–43.

[3] V.V. Batyrev and Yu. Tschinkel, Manin's conjecture for toric varieties. *J. Alg. Geom.* **7** (1998), 15–53.

[4] R. Beheshti, Lines on projective hypersurfaces. *J. Reine Angew. Math.* **592** (2006), 1–21.

[5] D.J. Bernstein, Enumerating solutions to $p(a) + q(b) = r(c) + s(d)$. *Math. Comp.* **70** (2001), 389–394.

[6] B.J. Birch, Forms in many variables. *Proc. Roy. Soc. Ser. A* **265** (1961), 245–263.

[7] E. Bombieri and J. Pila, The number of integral points on arcs and ovals. *Duke Math. J.* **59** (1989), 337–357.

[8] R. Brauer, A note on systems of homogeneous algebraic equations. *Bull. Amer. Math. Soc.* **51** (1945), 749–755.

[9] R. de la Bretèche, Sur le nombre de points de hauteur bornée d'une certaine surface cubique singulière. *Astérisque* **251** (1998), 51–77.

[10] R. de la Bretèche, Nombre de points de hauteur bornée sur les surfaces de del Pezzo de degré 5. *Duke Math. J.* **113** (2002), 421–464.

[11] R. de la Bretèche and T.D. Browning, Sums of arithmetic functions over values of binary forms. *Acta Arith.* **125** (2006), 291–304.

[12] R. de la Bretèche and T.D. Browning, On Manin's conjecture for singular del Pezzo surfaces of degree four, I. *Michigan Math. J.* **55** (2007), 51–80.

[13] R. de la Bretèche and T.D. Browning, On Manin's conjecture for singular del Pezzo surfaces of degree four, II. *Math. Proc. Camb. Phil. Soc.* **143** (2007), 579–605.

[14] R. de la Bretèche and T.D. Browning, Binary linear forms as sums of two squares. *Compositio Math.* **144** (2008), 1375–1402.

[15] R. de la Bretèche and T.D. Browning, Manin's conjecture for quartic del Pezzo surfaces with a conic fibration *Submitted*, 2008.

[16] R. de la Bretèche and T.D. Browning, Le problème des diviseurs pour des formes binaires de degré 4. *Submitted*, 2008.

[17] R. de la Bretèche and É. Fouvry, L'éclaté du plan projectif en quatre points dont deux conjugués. *J. Reine Angew. Math.* **576** (2004), 63–122.

[18] R. de la Bretèche and P. Swinnerton-Dyer, Fonction zêta des hauteurs associée à une certaine surface cubique. *Bull. Soc. Math. France* **135** (2007), 65–92.

[19] R. de la Bretèche, T.D. Browning and U. Derenthal, On Manin's conjecture for a certain singular cubic surface. *Ann. Sci. École Norm. Sup.* **40** (2007), 1–50.

[20] R. de la Bretèche, T.D. Browning and E. Peyre, On Manin's conjecture for a family of Châtelet surfaces. *In preparation.*

[21] T.D. Browning, Equal sums of two kth powers. *J. Number Theory* **96** (2002), 293–318.

[22] T.D. Browning, The density of rational points on a certain singular cubic surface. *J. Number Theory* **119**, (2006), 242–283.

[23] T.D. Browning, An overview of Manin's conjecture for del Pezzo surfaces. *Analytic number theory — A tribute to Gauss and Dirichlet*, Clay Math. Proceedings **7** (2007), 39–56.

[24] T.D. Browning, Counting rational points on cubic hypersurfaces. *Mathematika* **54** (2007), 93–112.

[25] T.D. Browning, Linear growth for Châtelet surfaces. *Math. Annalen*, to appear.

[26] T.D. Browning and U. Derenthal, Manin's conjecture for a quartic del Pezzo surface with $\mathbf{A}_4$ singularity. *Ann. Inst. Fourier* **59** (2009), 1231–1265.

[27] T.D. Browning and U. Derenthal, Manin's conjecture for a cubic surface with $\mathbf{D}_5$ singularity. *Int. Math. Res. Not.* **14** (2009), 2620–2647.

[28] T.D. Browning and R. Dietmann, Representation of integers by quadratic forms. *Proc. London Math. Soc.* **96** (2008), 389–416.

[29] T.D. Browning and R. Dietmann, Solubility of Fermat equations. *Quadratic forms — algebra, arithmetic and geometry*, 99–106, Contemporary Mathematics **493**, 2009.

[30] T.D. Browning and D.R. Heath-Brown, Counting rational points on hypersurfaces. *J. Reine Angew. Math.* **584** (2005), 83–115.

[31] T.D. Browning and D.R. Heath-Brown, Plane curves in boxes and equal sums of like powers. *Math. Zeit.* **251** (2005), 233–247.

[32] T.D. Browning and D.R. Heath-Brown, The density of rational points on non-singular hypersurfaces, I. *Bull. London Math. Soc.* **38** (2006), 401–410.

[33] T.D. Browning and D.R. Heath-Brown, Counting rational points on smooth hypersurfaces, II. *Proc. London Math. Soc.* **93** (2006), 273–303.

[34] T.D. Browning and D.R. Heath-Brown, Rational points on quartic hypersurfaces. *J. Reine Angew. Math.* **629** (2009), 37–88.

[35] T.D. Browning, R. Dietmann and P.D.T.A. Elliot, Least zero of a cubic form. *Submitted*, 2009.

[36] T.D. Browning, D.R. Heath-Brown and P. Salberger, Counting rational points on algebraic varieties. *Duke Math. J.* **132** (2006), 545–578.

[37] J.W. Bruce and C.T.C. Wall, On the classification of cubic surfaces. *J. London Math. Soc.* **19** (1979), 245–256.

[38] J.W.S. Cassels, Bounds for the least solutions of homogeneous quadratic equations. *Proc. Cambridge Philos. Soc.* **51** (1955), 262–264.

[39] J.W.S. Cassels, Addendum to the paper "Bounds for the least solutions of homogeneous quadratic equations". *Proc. Cambridge Philos. Soc.* **52** (1956), 604.

[40] J.W.S. Cassels, *Introduction to the geometry of numbers.* Springer-Verlag, 1971.

[41] A. Cayley, A memoir on cubic surfaces. *Phil. Trans. Roy. Soc.* **159** (1869), 231–326.

[42] A. Chambert-Loir and Yu. Tschinkel, On the distribution of points of bounded height on equivariant compactifications of vector groups. *Invent. Math.* **148** (2002), 421–452.

[43] J.-L. Colliot-Thélène and J.-J. Sansuc, Torseurs sous des groupes de type multiplicatif; applications à l'étude des points rationnels de certaines variétés algébriques. *C. R. Acad. Sci. Paris Sér. A-B* **282** (1976), 1113–1116.

[44] J.-L. Colliot-Thélène and J.-J. Sansuc, La descente sur les variétés rationnelles, II. *Duke Math. J.* **54** (1987), 375–492.

[45] J.-L. Colliot-Thélène, J.-J. Sansuc and P. Swinnerton-Dyer, Intersections of two quadrics and Châtelet surfaces, I. *J. Reine Angew. Math.* **373** (1987), 37–107.

[46] J.-L. Colliot-Thélène, J.-J. Sansuc and P. Swinnerton-Dyer, Intersections of two quadrics and Châtelet surfaces, II. *J. Reine Angew. Math.* **374** (1987), 72–168.

[47] D.F. Coray and M.A. Tsfasman, Arithmetic on singular Del Pezzo surfaces. *Proc. London Math. Soc.* **57** (1988), 25–87.

[48] S. Daniel, On the divisor-sum problem for binary forms. *J. Reine Angew. Math.* **507** (1999), 107–129.

[49] H. Davenport, Cubic forms in 16 variables. *Proc. Roy. Soc. A* **272** (1963), 285–303.

[50] H. Davenport, *Analytic methods for Diophantine equations and Diophantine inequalities.* 2nd ed., edited by T.D. Browning, Cambridge University Press, 2005.

[51] P. Deligne, La conjecture de Weil, I. *Inst. Hautes Études Sci. Publ. Math.* **43** (1974), 273–307.

[52] P. Deligne and N. Katz, Groupes de monodromie en géométrie algébrique, II. *SGA 7 II*, Lecture Notes in Math. **340**, Springer-Verlag, 1973.

[53] U. Derenthal, Singular del Pezzo surfaces whose universal torsors are hypersurfaces. *arXiv:math.AG/0604194*, 2006.

[54] U. Derenthal, Counting integral points on universal torsors. *Int. Math. Res. Not.* **14** (2009), 2648–2699.

[55] U. Derenthal and Yu. Tschinkel, Universal torsors over Del Pezzo surfaces and rational points. *Equidistribution in number theory, an introduction*, 169–196, NATO Science Series II **237**, Springer-Verlag, 2007.

[56] N. Elkies, On $A^4 + B^4 + C^4 = D^4$. *Math. Comp.* **51** (1988), 825–835.

[57] J. Ellenberg and A. Venkatesh, On uniform bounds for rational points on nonrational curves. *Int. Math. Res. Not.* **35** (2005), 2163–2181.

[58] G. Faltings, Endlichkeitssätze für abelsche Varietäten über Zahlkörpern. *Invent. Math.* **73** (1983), 349–366.

[59] É. Fouvry, Sur la hauteur des points d'une certaine surface cubique singulière. *Astérisque* **251** (1998), 31–49.

[60] J. Franke, Yu.I. Manin and Yu. Tschinkel, Rational points of bounded height on Fano varieties. *Invent. Math.* **95** (1989), 421–435.

[61] M. Fujiwara Upper bounds for the number of lattice points on hypersurfaces. *Number theory and combinatorics (Japan, 1984)*, 89–96, World Scientific Publishing Co., 1985.

[62] G. Greaves, On the divisor-sum problem for binary cubic forms. *Acta Arith.* **17** (1970), 1–28.

[63] C.R. Guo, On solvability of ternary quadratic forms. *Proc. London Math. Soc.* **70** (1995), 241–263.

[64] G.H. Hardy and J.E. Littlewood, A new solution of Waring's problem. *Quart J. Math.* **48** (1920), 272–293.

[65] J. Harris, *Algebraic geometry. A first course.* Springer-Verlag, 1992.

[66] R. Hartshorne, *Algebraic geometry.* Springer-Verlag, 1977.

[67] B. Hassett and Yu. Tschinkel, Universal torsors and Cox rings. *Arithmetic of higher-dimensional algebraic varieties (Palo Alto, CA, 2002)*, 149–173, Progr. Math. **226**, Birkhäuser, 2004.

[68] D.R. Heath-Brown, Diophantine approximation with square-free numbers. *Math. Zeit.* **187** (1984), 335–344.

[69] D.R. Heath-Brown, The density of zeros of forms for which weak approximation fails. *Math. Comp.* **59** (1992), 613–623.

[70] D.R. Heath-Brown, The density of rational points on non-singular hypersurfaces. *Proc. Indian Acad. Sci.* **104** (1994), 13–29.

[71] D.R. Heath-Brown, The density of rational points on cubic surfaces. *Acta Arith.* **79** (1997), 17–30.

[72] D.R. Heath-Brown, The circle method and diagonal cubic forms. *R. Soc. Lond. Philos. Trans. Ser. A* **356** (1998), 673–699.

[73] D.R. Heath-Brown, Counting rational points on cubic surfaces. *Astérisque* **251** (1998), 13–29.

[74] D.R. Heath-Brown, The density of rational points on curves and surfaces. *Annals of Math.* **155** (2002), 553–595.

[75] D.R. Heath-Brown, Linear relations amongst sums of two squares. *Number theory and algebraic geometry*, 133–176, Lond. Math. Soc. Lecture Note Ser. **303** Cambridge University Press, 2003.

[76] D.R. Heath-Brown, The density of rational points on Cayley's cubic surface. *Proceedings of the session in analytic number theory and Diophantine equations*, Bonner Math. Schriften **360**, 2003.

[77] D.R. Heath-Brown, Counting rational points on algebraic varieties. *Analytic number theory*, 51–95, Lecture Notes in Math. **1891**, Springer-Verlag, 2006.

[78] D.R. Heath-Brown, Cubic forms in 14 variables. *Invent. Math.* **170** (2007), 199–230.

[79] D.R. Heath-Brown and B.Z. Moroz, The density of rational points on the cubic surface $X_0^3 = X_1X_2X_3$. *Math. Proc. Camb. Soc.* **125** (1999), 385–395.

[80] W.V.D. Hodge and D. Pedoe, *Methods of algebraic geometry.* Vol. 2, Cambridge University Press, 1952.

[81] C. Hooley, On another sieve method and the numbers that are a sum of two hth powers. *Proc. London Math. Soc.* **43** (1981), 73–109.

[82] C. Hooley, On another sieve method and the numbers that are a sum of two hth powers, II. *J. Reine Angew. Math.* **475** (1996), 55–75.

[83] C. Hooley, On Hypothesis K^* in Waring's problem. *Sieve methods, exponential sums, and their applications in number theory (Cardiff, 1995)*, 175–185, London Math. Soc. Lecture Note Ser. **237**, Cambridge University Press, 1997.

[84] K. Ireland and M. Rosen, *A classical introduction to modern number theory.* 2nd ed., Springer-Verlag, 1990.

[85] M. Joyce, *Rational points on the* $\mathbf{E}_6$ *cubic surface.* Ph.D. thesis, Brown University, 2008.

[86] H. Knörrer, *Isolierte singularitäten von durchschnitten zweier quadriken.* Bonner Mathematische Schriften **117**, 1980.

[87] L.J. Lander and T.R. Parkin, Counterexample to Euler's conjecture on sums of like powers. *Bull. Amer. Math. Soc.* **72** (1966), 1079.

[88] S. Lang and A. Weil, Number of points of varieties in finite fields. *Amer. J. Math.* **76** (1954), 819–827.

[89] F.-S. Leung, *Manin's conjecture on a non-singular quartic del Pezzo surface.* D.Phil thesis, Oxford University, 2008.

[90] J. Lipman, Rational singularities, with applications to algebraic surfaces and unique factorization. *Inst. Hautes Études Sci. Publ. Math.* **36** (1969), 195–279.

[91] Yu.I. Manin, *Cubic forms.* 2nd ed., North-Holland Mathematical Library **4**, North-Holland Publishing Co., 1986.

[92] O. Marmon, The density of integral points on complete intersections. *Quart. J. Math.* **58** (2007), 1–25.

[93] J.-F. Mestre, Courbes elliptiques et formules explicites, *Séminaire de théorie des nombres (Paris 1981–82)*, 179–182, Progr. Math. **38**, Birkhäuser, 1983.

[94] L.J. Mordell, A remark on indeterminate equations in several variables. *J. London Math. Soc.* **12** (1937), 127–129.

[95] W. Narkiewicz, *Number theory.* World Scientific Publishing Co., 1983.

[96] E. Peyre, Hauteurs et nombres de Tamagawa sur les variétés de Fano. *Duke Math. J.* **79** (1995), 101–218.

[97] E. Peyre, Counting points on varieties using universal torsors. *Arithmetic of higher-dimensional algebraic varieties (Palo Alto, CA, 2002)*, 61–81, Progr. Math. **226**, Birkhäuser, 2004.

[98] E. Peyre and Yu. Tschinkel, Tamagawa numbers of diagonal cubic surfaces, numerical evidence. *Math. Comp.* **70** (2001), 367–387.

[99] E. Peyre and Yu. Tschinkel, Tamagawa numbers of diagonal cubic surfaces of higher rank. *Rational points on algebraic varieties*, 275–305, Progr. Math., **199**, Birkhäuser, 2001.

[100] J. Pila, Density of integral and rational points on varieties. *Astérisque* **228** (1995), 183–187.

[101] J. Pitman, Cubic inequalities. *J. London Math. Soc.* **43** 1968, 119–126.

[102] P. Salberger, Tamagawa measures on universal torsors and points of bounded height on Fano varieties. *Astérisque* **251** (1998), 91–258.

[103] P. Salberger, Counting rational points on hypersurfaces of low dimension. *Ann. Sci. École Norm. Sup.* **38** (2005), 93–115.

[104] P. Salberger, On the density of rational and integral points on algebraic varieties. *J. Reine Angew. Math.* **606** (2007), 123–147.

[105] P. Salberger, Rational points of bounded height on projective surfaces. *Math. Z.* **258** (2008), 805–826.

[106] P. Salberger, Counting rational points on projective varieties. *Submitted*, 2009.

[107] S.H. Schanuel, Heights in number fields. *Bull. Soc. Math. France* **107** (1979), 443–449.

[108] L. Schläfli, On the distribution of surfaces of the third order into species. *Phil. Trans. Roy. Soc.* **153** (1864), 193–247.

[109] B. Segre, A note on arithmetical properties of cubic surfaces. *J. London Math. Soc.* **18** (1943), 24–31.

[110] B. Segre, Sulle V_n contenenti più di $\infty^{n-k} S_k$, I. *Atti Accad. Naz. Lincei. Rend. Cl. Sci. Fis. Mat. Nat.*, **5** (1948), 193–197, II, *Atti Accad. Naz. Lincei. Rend. Cl. Sci. Fis. Mat. Nat.*, **5** (1948), 275–280.

[111] J.-P. Serre, Spécialisation des éléments de $\mathrm{Br}_2(\mathbb{Q}(T_1, \dots, T_n))$. *C. R. Acad. Sci. Paris* **311** (1990), 397–402.

[112] J.-P. Serre, *Lectures on the Mordell–Weil theorem.* 3rd ed., Friedr. Vieweg & Sohn, 1997.

[113] C.M. Skinner and T.D. Wooley, Sums of two kth powers. *J. Reine Angew. Math.* **462** (1995), 57–68.

[114] J.B. Slater and P. Swinnerton-Dyer, Counting points on cubic surfaces, I. *Astérisque* **251** (1998), 1–12.

[115] P. Swinnerton-Dyer, Diophantine equations: progress and problems. *Arithmetic of higher-dimensional algebraic varieties (Palo Alto, CA, 2002)*, 3–35, Progr. Math. **226**, Birkhäuser, 2004.

[116] P. Swinnerton-Dyer, Counting points on cubic surfaces, II. *Geometric methods in algebra and number theory*, 303–310, Progr. Math. **235**, Birkhäuser, 2005.

[117] G. Tenenbaum, *Introduction to analytic and probabilistic number theory.* Cambridge University Press, 1995.

[118] G. Terjanian, Un contre-exemple à une conjecture d'Artin. *C. R. Acad. Sci. Paris Sér. A–B* **262** (1966), A612.

[119] R.C. Vaughan and T.D. Wooley, A special case of Vinogradov's mean value theorem. *Acta Arithmetica* **79** (1997), 193–204.

[120] R.C. Vaughan and T.D. Wooley, Waring's problem: a survey. *Number theory for the millenium, Vol. III (Bennett et al., eds.)*, 301–340, A.K. Peters, 2002.

[121] A. Weil, Numbers of solutions of equations in finite fields. *Bull. Amer. Math. Soc.* **55** (1949), 497–508.

[122] T.D. Wooley, Large improvements in Waring's problem. *Annals of Math.* **135** (1992), 131–164.

[123] T.D. Wooley, Some remarks on Vinogradov's mean value theorem and Tarry's problem. *Monatsh. Math.* **122** (1996), 265–273.

[124] T.D. Wooley, On the local solubility of Diophantine systems. *Compositio Math.* **111** (1998), 149–165.

[125] T. D. Wooley, Diophantine problems in many variables: the role of additive number theory. *Topics in number theory (S. D. Ahlgren et al., eds.)*, 49–83, Kluwer Academic Publishers, 1999.

[126] F.L. Zak, Determinants of projective varieties and their degrees. *Algebraic transformation groups and algebraic varieties*, Encyclopaedia of Mathematical Sciences **132**, Springer-Verlag, 2004.

Index

BIRKHÄUSER

Progress in Mathematics (PM)

Edited by
Hyman Bass, University of Michigan, USA
Joseph Oesterlé, Institut Henri Poincaré, Université Paris VI, France
Alan Weinstein, University of California, Berkeley, USA

Progress in Mathematics is a series of books intended for professional mathematicians and scientists, encompassing all areas of pure mathematics. This distinguished series, which began in 1979, includes research level monographs, polished notes arising from seminars or lecture series, graduate level textbooks, and proceedings of focused and refereed conferences. It is designed as a vehicle for reporting ongoing research as well as expositions of particular subject areas.

PM 259: Capogna, L. / Danielli, D. / Pauls, S.D. / Tyson, J.T.
An Introduction to the Heisenberg Group and the Sub-Riemannian Isoperimetric Problem (2007)
ISBN 978-3-7643-8132-5

PM 258: Gan, W.T. / Kudla, S.S. / Tschinkel, Y. (Eds.). Eisenstein Series and Applications (2008).
ISBN 978-0-8176-4496-3

PM 257: Lakshmibai, V. / Littelmann, P. / Seshadri, C.S. (Eds.)
Schubert Varieties (due 2009)
ISBN 978-0-8176-4153-5

PM 256: Borthwick, D.
Spectral Theory of Infinite-Volume Hyperbolic Surfaces (2007). ISBN 978-0-8176-4524-3

PM 255: Kobayashi, T. / Schmid, W. / Yang, J.-H. (Eds.)
Representation Theory and Automorphic Forms (2007). ISBN 978-0-8176-4505-2

PM 254: Ma, X. / Marinescu, G.
Holomorphic Morse Inequalities and Bergman Kernels (2007)
Winner of the Ferran Sunyer i Balaguer Prize 2006.
ISBN 978-3-7643-8096-0

PM 253: Ginzburg, V. (Ed.)
Algebraic Geometry and Number Theory (2006).
ISBN 978-0-8176-4471-7

PM 252: Maeda, Y. / Michor, P. / Ochiai, T. / Yoshioka, A. (Eds.)
From Geometry to Quantum Mechanics (2007).
ISBN 978-0-8176-4512-8

PM 251: Boutet de Monvel, A. / Buchholz, D. / Iagolnitzer, D. / Moschella, U. (Eds.)
Rigorous Quantum Field Theory (2006). ISBN 978-3-7643-7433-4

PM 250: Unterberger, A.
The Fourfold Way in Real Analysis (2006). ISBN 978-3-7643-7544-7

PM 249: Kock, J. / Vainsencher, I.
An Invitation to Quantum Cohomology. Kontsevich's Formula for Rational Plane Curves (2006).
ISBN 978-0-8176-4456-7

PM 248: Bartholdi, L. / Ceccherini-Silberstein, T. / Smirnova-Nagnibeda, T. / Zuk, A.
(Eds.) Infinite Groups: Geometric, Combinatorial and Dynamical Aspects (2006)
ISBN 978-3-7643-7446-4

PM 247: Baues, H.-J.
The Algebra of Secondary Cohomology Operations (2006)
ISBN 978-3-7643-7448-8

PM 246: Dragomir, S. / Tomassini, G., Differential Geometry and Analysis on CR Manifolds (2006).
ISBN 978-0-8176-4388-1

PM 245: Fels, G. / Huckleberry, A.T. / Wolf, J.A.
Cycle Spaces of Flag Domains. A Complex Geometric Viewpoint (2006). ISBN 978-0-8176-4391-1

PM 244: Etingof, P. / Retakh, V. / Singer, I.M. (Eds.)
The Unity of Mathematics (2006)
ISBN 978-0-8176-4076-7

PM 243: Bernstein, J. / Hinich, V. / Melnikov, A. (Eds.)
Studies in Lie Theory (2006)
ISBN 0-8176-4342-3

PM 242: Dufour, J.-P. / Zung, N.T.
Poisson Structures and their Normal Forms (2005)
ISBN 978-3-7643-7334-4

PM 241: Seade, J.
On the Topology of Isolated Singularities in Analytic Spaces
Winner of the Ferran Sunyer i Balaguer Prize 2005
ISBN 978-3-7643-7322-1

PM 240: Ambrosetti, A. / Malchiodi, A. Perturbation Methods and Semilinear Elliptic Problems in R^n (2005). *Winner of the Ferran Sunyer i Balaguer Prize 2005.*
ISBN 978-3-7643-7321-4

PM 239: van der Geer, G. / Moonen, B.J.J. / Schoof, R. (Eds.)
Number Fields and Function Fields – Two Parallel Worlds (2005)
ISBN 978-0-8176-4397-3

PM 238: Sabadini, I. / Struppa, D.C. / Walnut, D.F. (Eds.)
Harmonic Analysis, Signal Processing, and Complexity (2005).
ISBN 978-0-8176-4358-4

PM 237: Kulish, P.P. / Manojlovic, N. / Samtleben, H. (Eds.)
Infinite Dimensional Algebras and Quantum Integrable Systems (2005). ISBN 978-3-7643-7215-6

PM 236: Hotta, R. / Takeuchi, K. / Tanisaki, T.
D-Modules, Perverse Sheaves, and Representation Theory (due 2007)
ISBN 978-0-8176-4363-8

MIX
Papier aus verantwortungsvollen Quellen
Paper from responsible sources
FSC® C105338

If you have any concerns about our products,
you can contact us on
ProductSafety@springernature.com

In case Publisher is established outside the EU,
the EU authorized representative is:
Springer Nature Customer Service Center GmbH
Europaplatz 3, 69115 Heidelberg, Germany

Printed by Libri Plureos GmbH
in Hamburg, Germany